FROM THE BIG BANG TO PLANET X

FROM THE BIG BANG TO PLANET X

TERENCE DICKINSON

THE 50 MOST-ASKED QUESTIONS ABOUT THE UNIVERSE...AND THEIR ANSWERS

CAMDEN HOUSE

Canadian Cataloguing in Publication Data

Dickinson, Terence
From the big bang to planet X : the 50 most-asked questions about the universe—and their answers

Includes bibliographical references and index.
ISBN 0-921820-71-2

1. Astronomy - Miscellanea. I. Title.

QB52.D5 1993 520'.2 C93-094507-7

Published by Camden House Publishing
(a division of Telemedia Communications Inc.)

Camden House Publishing
7 Queen Victoria Road
Camden East, Ontario K0K 1J0

Camden House Publishing
Box 766
Buffalo, New York 14240-0766

Trade distribution by: Firefly Books
250 Sparks Avenue
Willowdale, Ontario
Canada M2H 2S4

Box 1325
Ellicott Station
Buffalo, New York 14205

Design by: Linda J. Menyes

Illustrations by: Margo Stahl

Cover: The Andromeda Galaxy, a wheel-shaped assemblage of about 500 billion stars, 2.3 million light-years from Earth. Photograph by Tony Hallas and Daphne Mount.

Filmwork by:
Hadwen Graphics Limited, Ottawa, Ontario

Printed and bound in Canada by:
D.W. Friesen & Sons, Altona, Manitoba

To my brother Steve,
an inquisitive and passionate naturalist.

ACKNOWLEDGMENTS

One summer night in the late 1950s, a teenager sat alone at the back of a bus, returning home after an unusual experience. Two long, thin boxes lay on the seat beside him. The boxes held the tripod and tube of his telescope—a prized possession of the 16-year-old. He had just participated in his first public "star party." He and about 10 other members of the Royal Astronomical Society of Canada had set up their telescopes in a park, and hundreds of people had lined up for a look. Astronomy clubs all over the world do this, but it was a new experience for the young amateur astronomer—me.

I was brimming with enthusiasm about astronomy. I had read everything on the subject that I could get my hands on. But the conversations with the people who looked through my little telescope that night opened a new window for me. What I learned for the first time, and have had reinforced over and over again in the years since, is that certain aspects of the universe provoke deep curiosity among many people. Today, the list of "hot" astronomy subjects—and questions about them—is longer than it was back in 1959, but I have tried to include as much as I could in this book, while endeavouring to keep it concise and accessible.

Making it all come together was Linda Menyes, the talented designer of the book who made my work so much easier, along with technical illustrator Margo Stahl, who had some tough cosmology assignments, and proofreaders Christine Kulyk, Catherine DeLury and Laura Elston. Few writers ever get to work with an editor with such unreserved enthusiasm as Tracy Read. Tracy's perceptive advice improved the book in many ways. My thanks also to Alan Dyer, who commented on some early versions of the manuscript. And, finally, thanks to my best friend and wife Susan, possibly the finest copy editor and proofreader in the universe.

CONTENTS

INTRODUCTION

What came before the Big Bang? What is a black hole? Why do stars twinkle? Are UFOs real?

Who asks questions like these?

You do.

How do I know? I have heard them asked *many* times. For more than 30 years, I have talked about astronomy in planetariums and on radio, written about the cosmos in books, magazines and newspapers and taught astronomy at college. Year after year, again and again, I am asked many of the same questions—or variations of them. The 50 most-asked questions are the basis for this book.

If you have ever gazed in wonderment at a star-filled sky on a cool summer night in the country or had your curiosity aroused by a brief television news item about an astronomical discovery, then this book is for you. You can jump in anywhere; the book is designed for browsing. It provides more depth than television can—but never more than you can handle. And if you want to explore any

The Horsehead Nebula is part of a vast cloud of gas and dust in the constellation Orion. Every star in this and all other celestial photographs, no matter how bright or dim it appears, is a sun. Astronomers estimate that there are at least a billion trillion stars in the universe. Photograph by John Mirtle.

Our own galaxy, the Milky Way, as it might appear from just above the clouds of a hypothetical planet about 100,000 light-years away. Our sun, one of roughly 300 billion stars in the galaxy, is two-thirds of the way out from the centre to the edge. Illustration by Adolf Schaller.

of the subjects further, I have provided suggestions for in-depth reading for every question. Before you start, though, the following diversion should make the browsing easier.

Picture the interior of Toronto's huge retractable-roof stadium, the SkyDome, several hours before a baseball game. The seats are empty, and the grounds crew has taken a coffee break. At home plate rests the sun, the size of a baseball. The first four planets —Mercury, Venus, Earth and Mars—each about the size of the ball in a ballpoint pen, are, respectively, 1/8, 1/5, 1/3 and 1/2 of the way to the pitcher's mound. A pea near second base is Jupiter. In shallow centre field is Saturn, a smaller pea. Uranus, the size of this letter O, is at the outfield fence. Neptune, another O, is well back in the outfield seats, while Pluto, the smallest planet and a mere grain of sand in our model, runs along the track that carries the dome's movable sections.

Between Mars and Jupiter, just inside the bases, are millions of bits of dust—these are the asteroids. Ranging well beyond Pluto, out past the city limits, are trillions of comets, all so small that a microscope would be needed to see them. Back inside the dome, down in the artificial turf somewhere between home plate and the pitcher's mound, is the farthest humans have travelled: one thumb width, from Earth to the moon.

The utility of our model does not stop there. The nearest star, Alpha Centauri, can be included too, represented by another baseball, at another home plate—in the Houston Astrodome.

Such perspectives make it clear why sending robot spacecraft to the planets orbiting our sun (which we have done, with the exception of Pluto) is completely different from dispatching vehicles to explore planets of other stars. Interstellar travel is a quantum leap in distance and technological difficulty, not to mention the travel time. A robot spacecraft trekking to Alpha Centauri at the same speed as today's fastest spacecraft would take 30,000 years to get there. Even travelling at 1 percent of the velocity of light, a speed enormously faster than any current technology can muster, the voyage would still take 430 years.

Alpha Centauri's distance is 41 trillion kilometres. Astronomers use a more convenient measure: 4.3 light-years. One light-year, the distance light travels in one year, is 9.46 trillion kilometres. The farthest star seen with the naked eye on a dark night is about 4,000 light-years from Earth. That's still well within our galaxy, the Milky Way, which is a flat disc-shaped city of several hundred billion stars about 85,000 light-years across.

The nearest galaxy like our own, the Andromeda Galaxy, is 2.2 million light-years away. There are an estimated 100 billion similar galaxies in the known universe. The most remote of them, more than 10 billion light-years from us, show up as faint oval smudges on photographs taken with the Hubble Space Telescope. If scaled-down versions of the Milky Way Galaxy and the Andromeda Galaxy were placed at either side of the sprawling city of Los

Angeles, the distant galaxies seen by Hubble would be farther away than the moon.

This immensity, the mind-stretching vastness of the universe, means that it can *never* be fully explored in all its detail. The universe's estimated cargo of a billion trillion stars is so overwhelming that just trying to count them, at the rate of one per second, would require 500 billion human life spans. It would be the equivalent of counting all the grains of sand on all the Earth's beaches.

Yet as Einstein once said: "The most incomprehensible thing about the universe is that it is comprehensible." We *do* know an enormous amount about the cosmos and our place in it—just enough to begin to ask the big questions.

Terence Dickinson

ASTRONOMICAL TERMS USED THROUGHOUT THIS BOOK

A ***planet*** is a body that orbits a star, although not all objects which orbit stars are classed as planets. Astronomers consider a diameter of 1,000 kilometres to be the minimum size for a planet. Planets can be mostly rock, like Earth; mostly ice, like Pluto; or mostly gas, like Jupiter. Nine planets orbit the sun.

The ***solar system*** includes everything controlled by the sun's gravity; that is, all objects that orbit the sun—the nine planets and their companion moons, plus comets, asteroids and other debris.

A ***star*** is an object more than 25,000 times the mass of Earth. Such a body produces, through its own gravity, so much pressure at its core that it ignites itself. The core "burns" with a thermonuclear (hydrogen-bomb-type) fire that releases prodigious amounts of radiant energy. The sun, for example, is 333,000 times the mass of Earth. Stars range from slightly less than one-tenth to more than 100 times the mass of our sun.

Galaxies are vast congregations of stars held in a group by their mutual gravity. Galaxies contain anywhere from a few million to a few trillion stars. The sun is located in the Milky Way Galaxy, which has about 300 billion stars.

The ***universe*** is all that we know exists: all the gal-

Galaxies such as these, each one a great city of billions of stars, are detected by large telescopes in every direction. At one time, astronomers thought the galaxies were randomly distributed in space, but recent surveys have confirmed that almost all galaxies are members of clusters shaped like giant pancakes or ribbons, each with hundreds or thousands of galaxies. Between the clusters are enormous voids, sometimes hundreds of millions of light-years wide. Palomar Observatory photograph.

axies and everything else which astronomers have ever detected or ever will detect. This definition used to be unambiguous, but in recent years, cosmologists have theorized that countless other universes might exist in addition to the one we perceive. Other universes cannot be observed, but their existence is impossible to disprove. The term universe as used in this book refers to the *observable* universe, excluding any hypothetical "other" universes. When there is the possibility for confusion, the observable universe is called *our* universe.

A ***light-year*** is a unit of distance measurement—the distance light travels in a year at its velocity of 300,000 kilometres per second. One light-year is 9.46 trillion kilometres. Light-years are the standard units used to describe distances between stars and galaxies. Other units derived from the same concept, such as light-minutes, are used occasionally (but *never* light-centuries).

THE UNIVERSE

The Whirlpool Galaxy, about 25 million light-years from Earth, is roughly the same size as our home galaxy, the Milky Way. Several hundred million years ago, the Whirlpool was sideswiped by the smaller galaxy that, from our point of view, seems to be attached to the end of one of its spiral arms. Lick Observatory photograph.

Humans have a deep desire to know their roots. In the broadest context, this desire is expressed as a longing to understand how the universe itself came to exist. Just a few years ago, it appeared as though the answer might be forever out of reach. But as researchers pull aside one seemingly impenetrable veil after another, the origin and evolution of the universe is beginning to come into focus. Or, perhaps more accurately, we *think* it is.

Cosmology is the branch of science devoted to grappling with the universe's origin and destiny. As we approach the 21st century, cosmology's gospel is the Big Bang theory (see Question 1). Nothing else in astronomy has captured the interest of so many people as the Big Bang. In one huge brush stroke, it provides a beginning to the universe and several scenarios for its ultimate fate.

Powerful and profound, yet mystical and dramatic, the Big Bang theory is now firmly in place as the mind's blueprint of the universe. Today, no other

aspect of astronomy receives more attention. Hence that is where our Q&A begins. In this chapter, I have tried to provide both an introduction to cosmology for those new to the subject and the latest theories and discoveries of interest to readers already familiar with the basic idea of the Big Bang.

1 WHAT EXACTLY IS THE BIG BANG THEORY?

The Big Bang theory is the best explanation we have for the origin and evolution of the universe. It may be wrong. It may even seem childishly naïve a century from now. But since the mid-1960s, it has been the theory which best fits the universe that we observe around us.

What we know with certainty about the nature of the universe is that it is expanding—the galaxies are moving away from each other. By measuring the rate at which the galaxies are receding, and thus the rate at which the universe is expanding, astronomers can estimate how long the expansion has been going on. Somewhere between 10 billion and 20 billion years ago, they say, the universe must have been born in a genesis fireball that propelled the subsequent—and ongoing—expansion.

At time zero, the creation instant, the universe may have been smaller than the nucleus of an atom, an embryo that would ultimately yield the universe we see today—complete with people, planets and stars. One-hundredth of a second after time zero, the universe was a 100-billion-degree inferno of dashing subatomic particles. At that moment, the universe was no larger in volume than Earth yet contained all the basic ingredients for a billion trillion stars and probably as many planets.

As the hot primal universe expanded, it cooled enough for the subatomic particles to unite into atoms of hydrogen and helium. Further expansion over the next one billion to two billion years allowed the cooled hydrogen and helium and a smattering of other light elements to collect into huge clouds, many of which collapsed under their own gravity to form the galaxies with their cargoes of stars and planets. On at least one of those planets, life emerged, and a species evolved that is now seeking to know its own origins.

That, in a vastly abbreviated form, is the Big Bang theory.

2 HOW DO WE KNOW THE UNIVERSE IS EXPANDING?

On small scales—for instance, within the solar system or even within our galaxy, the Milky Way—the universe's expansion has no effect. It is overridden by the gravitational interplay among planets, stars and star clusters. Thus Earth itself is not expanding, nor are the planets and stars in our galaxy speeding away from each other. Within our galaxy, some stars are receding, some approaching; it all averages out.

Even the nearest galaxies, such as the Andromeda Galaxy, are not rushing away from us. It is only on the scale of galaxy clusters that the universe's overall expansion is evident. The group of 30 or so galaxies to which the Milky Way Galaxy belongs, called the Local Group, is like an island surrounded at varying distances by other galaxy clusters all receding from us. For example, the Hercules galaxy cluster, a vast collection of at least 20,000 galaxies, is receding from the Local Group at 40,000,000 kilometres per hour. Another island of galaxies, called the Hydra galaxy cluster, is rushing away at 250,000,000 kilometres per hour. That's nearly 3 percent of the velocity of light.

The most remote galaxies observed to date are flying away from us at the astounding rate of 95 percent of the speed of light. Astronomers harvest these incredible velocities by using a spectroscope attached to a telescope to spread the light from a galaxy into a spectrum. Lines in the spectrum identify certain common gases in the galaxy's stars and gas clouds. The researchers know where the lines should be for objects at rest. But receding objects always show the lines shifted toward the red end of the spectrum. This redshift, like the drop in pitch of a receding train horn, is an accurate measure of the recessional velocity.

In 1928, American physicist Howard Percy Robertson noticed that the redshifts of dimmer and pre-

COSMIC VIEW OF THE UNIVERSE

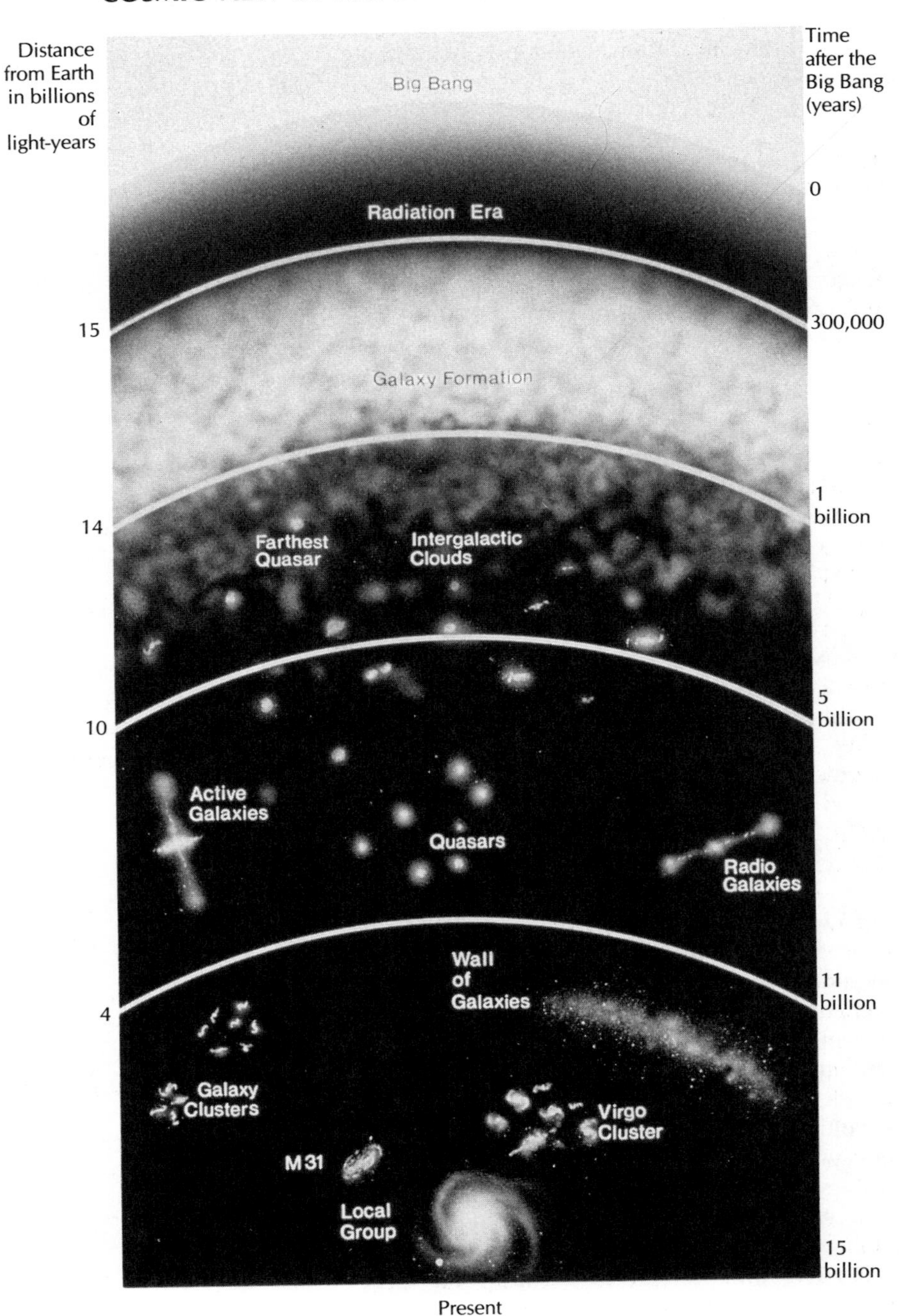

sumably more distant galaxies seem to be much larger than the redshifts of bright, obviously nearby galaxies. This is now regarded as a rule throughout the universe: *the faster a galaxy is receding, the farther away it is*. As an analogy, imagine how raisins in a rising blob of raisin-bread dough all move away from each other, and at the same time, the farther the raisins are apart, the faster they separate from each other.

Not only does this relationship show that the universe is expanding, but it allows astronomers to estimate the relative distances to galaxies using redshift alone, which makes it a very powerful tool.

3 *WHAT EVIDENCE IS THERE THAT THE BIG BANG REALLY HAPPENED?*

About 15 billion years ago, our universe may have started from a point no bigger than a proton. That crucible of creation is where and when space and time were born. As time grew from instant zero, space grew from size zero. As space expanded, there was more room in the dense, intensely hot sea of energy, and a steady cooling ensued. Eventually, space cooled enough for protons, neutrons and electrons to combine to form atoms. This happened about 300,000 years after the Big Bang.

Once matter in the universe was collected into atoms, light and other forms of radiation could begin to flow through space without being absorbed by jostling subatomic particles. That radiation is still moving through the universe today—weak, but detectable. Because the expansion of the universe over the past 15 billion years or so has cooled it to just three Celsius degrees above absolute zero, this radiation is called the three-degree cosmic background radiation.

The background radiation is the oldest thing astronomers can observe, and it carries with it the message of what the universe was like just 300,000 years after the Big Bang. Although its existence had been predicted, the background radiation was discovered quite by accident in 1965 during tests of a Bell Telephone Laboratories experimental microwave receiver. This ancient relic of our own origin

The universe as we perceive it is a window into the distant past. We see the universe as it was, not as it is. Because light from remote galaxies takes billions of years to reach us, the images our telescopes receive show us what was happening billions of years ago, a time when the universe was smaller and galaxies were younger and filled with energetic, newly formed stars. Galaxies were also closer and collided more readily with each other, producing quasars, the eruptive cores of galaxies in collision. The observation of galaxies in formation, one to two billion years after the Big Bang, is a realm that is just beginning to be probed. Radiant energy reaching us from farthest away, and thus furthest back in time, is called the cosmic background radiation —thought to be the first visible energy released after the Big Bang. Time and distance figures are very rough approximations. Illustration by Dana Berry, courtesy Space Telescope Science Institute.

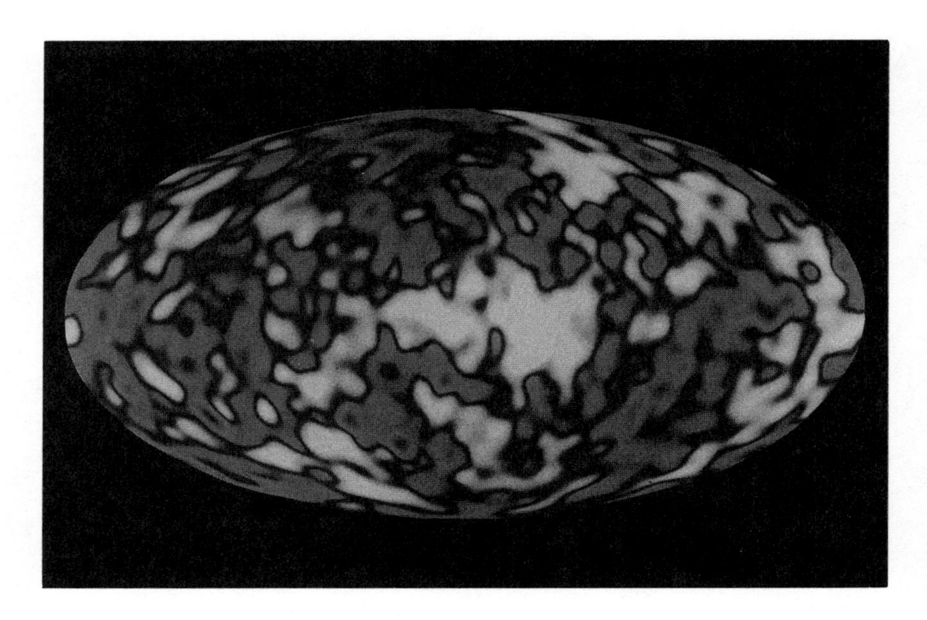

In 1992, NASA's Cosmic Background Explorer Satellite (COBE) provided this all-sky map of the background radiation, showing for the first time slight temperature variations that existed 300,000 years after the Big Bang. Although the temperature differences amount to only 3/100,000 of one Celsius degree, they were predicted and are believed to be the precursors to the galaxy clusters that would arise billions of years later. The COBE findings are widely regarded as supporting standard Big Bang cosmology. NASA photograph.

is now considered the best proof we have that the Big Bang actually happened.

In April 1992, astronomers announced that NASA's Cosmic Background Explorer satellite had detected slight variations in the temperature of the background radiation reaching us from different directions. The variations are tiny, just 3/100,000 of one Celsius degree from place to place, but they indicate that half a million years after the Big Bang, matter was not spread evenly over the universe. The slightly denser regions appear to be a bit cooler than the less dense regions.

This discovery caused a sensation, because it showed for the first time that matter had started to collect in patches even back near the beginning of time. Astronomers believe that those tiny density ripples were the seeds for vast clusters of galaxies which would arise one billion to two billion years later. These enormous galaxy clusters, sometimes made up of thousands of galaxies, are seen everywhere in the universe. The clusters are separated by voids hundreds of millions of light-years across that are inhabited by few, if any, galaxies. Had these ripples not been found, the emergence of galaxies, and especially galaxy clusters, from the Big Bang would have been tough to explain, and the Big Bang theory might have been in serious trouble. For now, the Big Bang is still the best genesis theory we have.

4 IF ALL THE GALAXIES ARE MOVING AWAY FROM US, DOES THAT MEAN WE ARE AT THE CENTRE OF THE UNIVERSE?

The universe would look essentially the same regardless of the galaxy from which it was observed. Every galaxy sees all the other galaxies hurtling away from it. Astronomers use the inflating-balloon demonstration to illustrate this point.

Stick small paper dots to the surface of an uninflated balloon. Each dot represents a galaxy. As the balloon is inflated, the dot galaxies retreat from each other. Each dot galaxy "observes" the same thing. There is no centre, nor is there an edge. The balloon's surface is a two-dimensional representation of the three-dimensional universe.

However, there is another element in the equation that does, in one sense, put us at the centre. We are the present, and everything else we see is in the past. We see the universe as it was, not as it is. A star 75 light-years away is seen as it was a human lifetime ago, because the light from the star has taken 75 years to reach Earth. But that's small-scale stuff. Where the effect begins to count is from one galaxy to another. Light reaching us from a galaxy five billion light-years away is five billion years old. We see the galaxy as it was five billion years ago, before Earth even existed.

What this means is that we perceive the universe as if from the centre of an onion, with each layer outward from the centre representing an increasingly earlier epoch. If we could look far enough, we might be able to see all the way back to the beginning of time. And that, in fact, is what is happening when astronomers examine the background radiation—at least, it is as close to the beginning as we will ever get. (The background radiation, as explained in Question 3, is the remnant of the first radiation emitted by the hot, young universe 300,000 years after its birth.)

The background radiation has reached us from near the beginning of time and from the edge of

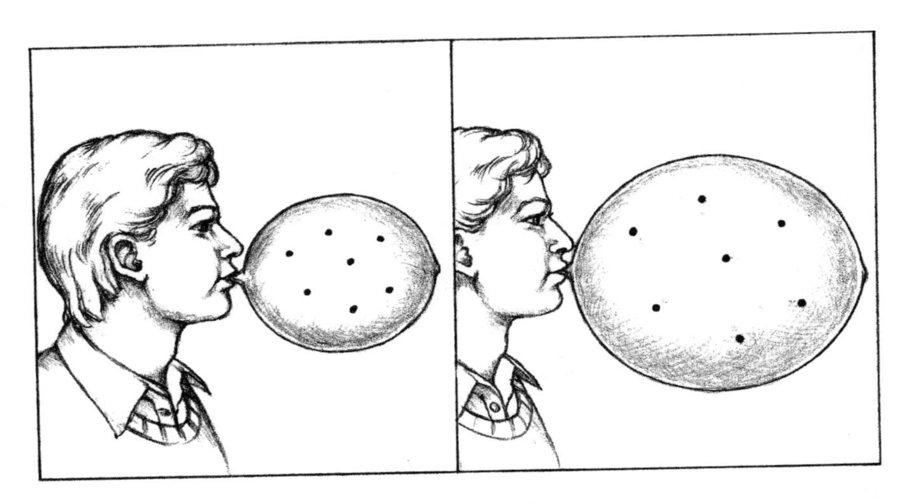

The expansion of the universe can be demonstrated in a two-dimensional analogy by sticking paper dots to the surface of a balloon, then inflating it. The galaxies themselves do not expand, but space does, as represented by the stretching balloon surface. Thus the galaxies move away from each other. Note that the farther apart the galaxies are, the faster they retreat from each other as the balloon inflates. This, too, reflects what happens in the real universe.

space as we perceive it. Nothing more ancient can be observed because its light was absorbed in the subatomic-particle chaos of the primordial universe. The background radiation is, in effect, the edge of our universe. It appears all around us, because the universe is all around us.

Thus from the point of view of what we observe, we *are* at the centre of the universe; the edge is the beginning, the centre is the present. But this is an illusion, a time-machine effect caused by the finite speed of light that brings us tales of what was, not what is. Moreover, observers on other galaxies would see the same thing: concentric spheres of the past enveloping them as they do us. To another civilization, our galaxy is seen as it was billions of years ago. We are the past in somebody else's present.

5 IF THE UNIVERSE IS EXPANDING, WHAT IS IT EXPANDING INTO?

It is best to try to avoid thinking of the Big Bang as an explosion, even though it is usually described that way. The problem with the word explosion is this: it conjures an image in the mind's eye that the Big Bang should be imagined from the outside, with the universe going "boom" and expanding "into" something. But it was not an explosion *into* space; rather, it was a simultaneous creation and expansion *of* space. We live in a universe where everything

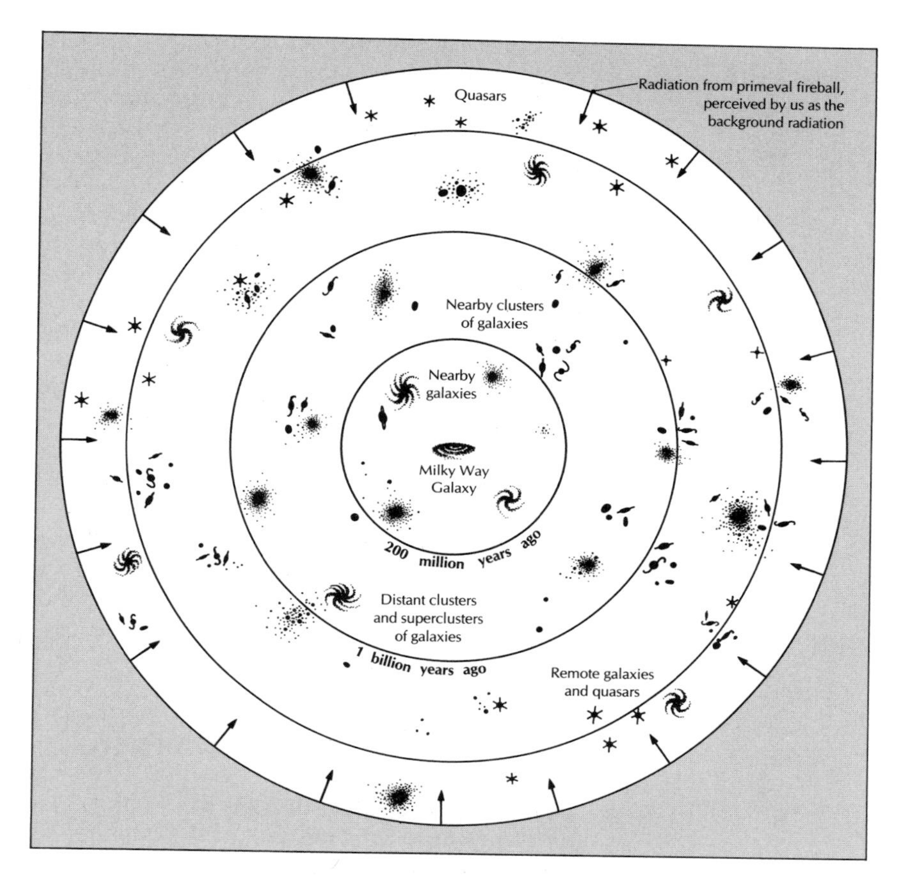

we can measure, observe or perhaps ever know is *inside*. There is no perceivable outside.

Time and space were created along with energy and matter in the Big Bang. We can observe only within that framework. Outside our universe is a realm that exists in another space-time regime—another dimension unconnected to our universe. If other universes exist (and cosmologists say there is no reason why they can't), they are sealed off from us by a barrier something like that which prevents us from visiting our own past. Therefore, no matter how large our universe becomes through its own expansion, it can never come into contact with another universe.

Although theoretical cosmologists say there is no conceivable way that we could escape from our universe or observe another one, I am often asked what

Whole-universe diagram: Our view of the universe is like penetrating the sequential layers of an onion, peering ever deeper into the past. The edge of the universe—the background radiation—is actually the beginning, the first release of energy after the Big Bang. Quasars, the most distant individual objects so far detected, can be seen out to at least 12 billion light-years.

other universes might be like IF we could visit them. Because there are an infinite number of ways that fundamentals such as time, space, gravity, matter and energy can be related to one another, each universe would have its own rule book formulated by the conditions which existed when that universe was created. The most likely fate for a visitor from our universe would be instant disintegration as our molecules suddenly came under the power of a new rule book for how matter behaves. If the visitor somehow survived the transition, the other universe might well be unrecognizable to our senses. Parallel universes (where everything is the same as our universe) are probably just wishful thinking.

There has been speculation that our universe might be connected to other universes by wormholes or by space-warping tunnels which might exist inside black holes (see Question 22), but this is not based on any reasonable extrapolation of what we now know. It's just a guess. In any case, if we did get outside, our universe would be as invisible to us as other universes are now.

6 *WHAT CAME BEFORE THE BIG BANG?*

Until recently, this question has been simply dismissed as unanswerable. But in the late 1970s, cosmology—the investigation of the origin and destiny of the universe—joined with a powerful ally, quantum mechanics—the study of the nature and behaviour of subatomic particles—to create the new science of quantum cosmology.

Giant particle accelerators such as Fermilab, near Chicago, and the European Centre for Nuclear Research, in Switzerland, have revealed how matter behaves at the enormous temperatures and pressures that existed in the primal fireball during the very early moments of the universe. Using this information, theorists have calculated what happened right back to the first millionth of a trillionth of a trillionth of a second after the genesis instant, while still remaining within the boundaries of known principles of physics.

During the 1980s, several breakthroughs in quantum cosmology allowed theorists to begin to specu-

Nineteenth-century woodcut by French astronomy popularizer Camille Flammarion shows an astronomer peering behind the visible universe to see the real workings of the cosmos. In a quaint yet compelling way, this widely reproduced illustration has come to represent the ultimate quest for the cosmological Holy Grail—a comprehensive understanding of the origin and destiny of the universe.

late about how the Big Bang happened and what came before. One concept favoured by many researchers in this field offers the seemingly fanciful hypothesis that our universe was created from nothing. Even more outlandish is the corollary: Our universe may be one of countless universes that have materialized out of pure nothingness.

In our everyday experience, far from the arcane realm of subatomic particles, nothingness is a pure vacuum—emptiness, the epitome of changelessness and inaction. But in the world of quantum physics, a vacuum is something outside classical space and time as we know it. It is a miniature universe quietly bubbling with its own inherently unstable energy.

On a subatomic scale, vacuums are alive with entities called virtual particles, which come dancing into existence only to vanish back into the froth of the quantum vacuum an instant later. This process, called quantum fluctuation, is so fundamental that physicists think it must occur everywhere, inside or outside our universe.

But the key point is that quantum mechanics allows for occasional random glitches which could keep a quantum fluctuation from vanishing. Once this happens, the rogue quantum fluctuation would tap into the immense latent energy in the vacuum and, in less than a trillionth of a trillionth of a second, be powered up with a universe's worth of energy such as the Big Bang possessed.

However, in the bizarre realm of quantum cosmology, no energy is stolen from the primordial vacuum; it is simply in a different form than it was before. All the matter in our universe consists of positive energy, which is apparently balanced by the negative energy of gravity. The total energy of the universe, therefore, could be exactly zero. As quantum cosmologist Alan Guth, one of the developers of this concept, once remarked: "Our universe seems to be the ultimate free lunch."

Furthermore, new universes could be aborning at any time, from inside our universe or outside. Quantum cosmology strongly implies that ours cannot be the only universe. We may be but a speck in an infinite universe of universes with no beginning and no end.

If this sounds like merely an invention by theorists to answer the thorny problem of genesis, all I can say is that I have a tough time with this stuff as well. But my difficulty is not in accepting it but in keeping a current understanding of it. Quantum physics deals exclusively with times, places, sizes and conditions utterly alien to everyday experience. However, it is not as if these are renegade ideas. They are embraced by leading theoretical cosmologists such as Stephen Hawking and have stood the test of more than a decade of debate. But perhaps all these notions will seem rather quaint by the 22nd century.

7 *ARE THERE THEORIES APART FROM THE BIG BANG?*

In the 1950s, the steady state theory shared centrestage with the Big Bang. According to steady state cosmology, the universe had no beginning, always existed and would continue forever, evolving as stars are born and die but overall remaining essentially the same. This theory accounts for the expansion of the universe and the creation of matter to fill it, but it fails to explain phenomena discovered since the mid-1960s, such as the background radiation, that are now pillars of the Big Bang theory.*

Physicist Dennis Sciama, one of the steady state's architects, mourned the theory's demise in a 1968 article. "The steady state theory is more beautiful and elegant than the real universe," he lamented. "Its loss has been a great sadness to me."

A more recent alternative to the Big Bang, called the plasma universe theory, is supported by a small minority of astronomers. This theory suggests that the universe has always existed and that its structure is dominated by electricity and magnetism, rather than gravity. While a state of matter called plasma—where atoms are separated into roaming electrons and protons—is rare on Earth, it occurs naturally in the near vacuum of space, and that plasma is alive with electric and magnetic fields.

On very large scales, according to plasma universe proponents, the universe's electric and magnetic fields organize matter into galaxies and stars. This may not be as farfetched as it sounds. Something as close to us as the aurora borealis is a plasma phenomenon.

A major strength of the plasma universe theory, say its supporters, is that it accounts for the vast billion-light-year-long ribbons of galaxy clusters which apparently pervade the cosmos. The Big Bang theory does too, but not with the same elegance. Plasma that occurs naturally in space would, by its nature, form vast filaments which could array the galaxies like strings of Christmas lights.

Proponents of the plasma universe theory say that it also offers an explanation for the dark-matter enigma. The dark-matter problem (see Question 11) stems from observations that galaxies move as if they are being affected by the gravitational attraction of far more matter than appears to be there. Despite a decade of investigation, the dark matter—if it exists—remains elusive.

Even though the plasma universe theory seems to address many cosmological problems, it is a theory waiting to be supported by observations, and few Big Bang advocates have converted. The majority

There are three possible fates for our universe: continuous expansion (top); a balanced scenario, where the universe has the exact mass density to bring the expansion to a halt but not reverse it (centre); and eventual collapse into a big crunch (bottom). The current stage of our universe's evolution is represented by the second box in each case. Astronomers consider the top and centre possibilities more likely than the bottom one, but the universe's destiny remains very much an unanswered question.

opinion is summed up by Princeton University cosmologist James Peebles: "So much evidence has accumulated in support of the Big Bang, it's still the best theory we have."

**Ironically, it was one of the strongest advocates of the steady state theory, cosmologist Fred Hoyle, who coined the term "Big Bang" during a series of astronomy talks he gave on BBC radio in 1950.*

8 HOW WILL OUR UNIVERSE END?

Some say the world will end in fire,
Some say in ice.

So wrote poet Robert Frost after hearing a lecture on the fate of the universe by Harvard University astronomer Harlow Shapley. Shapley had explained that there were only two possibilities: Either the universe would continue to expand forever, ultimately cooling to absolute zero as the last stars winked out, or the combined gravitational pull of all the galaxies and other matter in the universe would bring the expansion to a halt, followed by a collapse into a final big crunch. Shapley stated that nobody knew what fate the future might hold.

More than half a century later, we still do not know the universe's destiny, but one more option has emerged. The universe could be perfectly balanced so that its mass is precisely right to arrest the expansion eventually but never quite reverse it. Shapley hadn't considered such a possibility, because it seemed to him incredibly unlikely that nature could serve up a universe with such specific ingredients. But that is, in fact, what some cosmologists now suspect. From a theoretical standpoint, a balanced universe is mathematically the most satisfying of the three scenarios.

However, we are still a long way from predicting our universe's future conclusively. We do not even know how old it is, although estimates range from 9 billion to 24 billion years. To establish both the universe's present age and its fate, three crucial numbers are required. The first number is the rate at which the universe is expanding. By measuring the redshifts of remote galaxies and estimating their dis-

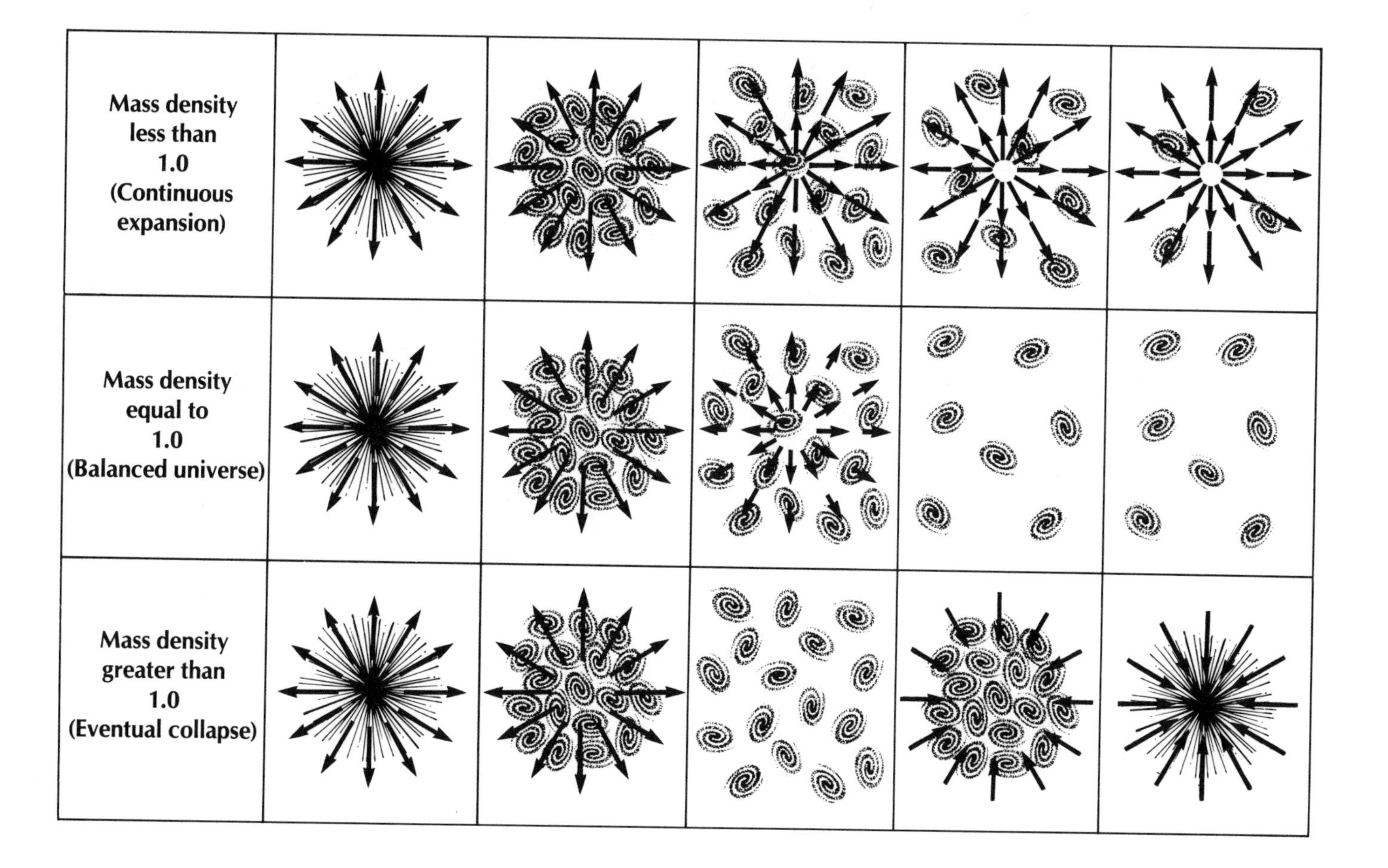

Mass density less than 1.0 (Continuous expansion)
Mass density equal to 1.0 (Balanced universe)
Mass density greater than 1.0 (Eventual collapse)

tances, astronomers have been able to approximate this number. It is somewhere between 40 and 90 kilometres per second per megaparsec of space. (A megaparsec, 3.2 million light-years, is a standard unit used by cosmologists.) The faster the rate at which the universe is expanding, the younger it is.

Moreover, the universe has mass, and the gravitational pull generated by that mass would slow the expansion rate over time. To find out how much deceleration has occurred and how much will occur in the future, we need the second number, called the mass density of the universe. For a balanced universe, the mass density is 1.0; that is, just enough to halt the expansion but not enough to reverse it. If the mass density is more than 1.0, the universe will eventually reverse and collapse on itself—Robert Frost's "end in fire." If it is less than 1.0, the universe will expand forever—Frost's ice destiny.

All the galaxies with all their stars and nebulas add up to an estimated mass density of less than 0.1. Yet for more than a generation, astronomers have measured the gravitational effect of some invisible stuff known as the dark matter (see Question 11) that "weighs" substantially more than the visible matter. There is no question that the dark matter exists, although its nature remains enigmatic. But when its weight is added to the visible matter, the mass density is at least 0.3, perhaps as high as 1.0.

The third number is a mysterious factor called the cosmological constant, which—if it exists—is a manifestation of the energy of empty space. (We discussed this in Question 6, when we referred to a vacuum as a storehouse of latent energy on a subatomic level.) As Einstein showed, mass and energy are equivalent. If the latent energy of space somehow acts as mass, it could add to the universe's mass density. At present, there is no evidence that the cosmological constant exists, but many theorists say a balanced universe (one with a mass density of 1.0) is the best fit to the current versions of the Big Bang theory. If all the ordinary matter and the dark matter in the universe add up to a mass density of less than 1.0, the only way to get a balanced universe is with a cosmological constant.

Theoreticians have developed mathematical models that show how different values for the three crucial numbers yield different ages and destinies for our universe. For example, if the rate of expansion is 45, the mass density is 0.3 and there is no cosmological constant, then the universe is 18 billion years old and will continue expanding forever. Raising the expansion rate to 75, the age drops to 11 billion years, but the fate remains the same. However, adding a cosmological constant to bump the mass density to 1.0, the present age would be 15 billion years with an expansion rate of 75 and 24 billion years with an expansion rate of 45.

Some of the models deliver a universe that is younger than 15 billion years, which is the age of the oldest known stars. Clearly, this is impossible; the universe can't be younger than the stars within it. Either the method used to estimate the ages of the stars is flawed or the models do not reflect reality. Yet many researchers are finding that an expansion rate of 75 to 85—which translates into a young universe in most models—best fits their observations.

The age/destiny question is far from settled. Stellar astronomers, cosmologists and theorists continue to debate the issue politely in scientific journals. There are, however, a few cosmology experts who have not spoken to each other for years, such is the heat of disagreement among them about the values for the three Holy Grail numbers that will eventually be used to define the evolution and fate of our universe.

9 *IF OUR UNIVERSE ENDS BY COLLAPSING, COULD A NEW UNIVERSE BE BORN?*

A popular theory from the 1950s, called the oscillating universe, began with the assumption that sometime in the distant future, the expanding universe would cease to expand and would begin to contract, eventually collapsing on itself in a big crunch, the reverse of the Big Bang. The oscillating universe concept proposed that the force of the implosion would then create another Big Bang—a new universe born from the ashes of the old.

The theory fell out of favour during the 1970s because theorists could not develop a reasonable mechanism to explain how the new universe would emerge. Moreover, evidence was piling up which indicated that the universe was not destined to collapse. Nevertheless, in 1990, University of Alberta astrophysicist Werner Israel retrieved the oscillating universe idea from the theoretical trash can by taking a new approach. He focused on the idea that black holes, the last objects existing before the final crush of a collapsing universe, would hold the key to its next evolution. (For background material on black holes, see Question 20.)

Israel's calculations suggest that the mass of the core of a black hole can increase to infinity as it absorbs and amplifies gravity waves. However, the inflated mass is a phenomenon limited to the core of a black hole, not the part in contact with our universe, so its consequences would remain invisible. As with the hydrogen-bomb energy contained within ordinary, harmless hydrogen atoms, specific conditions of immense pressure would be required for the black-hole core mass to become apparent. It emerges only when—and if—the universe collapses down to a dense stew of merging black holes.

"Then, when the black holes merge," explains Israel, "there are no more outside observers; everything is inside a black hole and is affected by it for the first time." This results in the sudden release of more than enough energy for another Big Bang. The newly created universe would be much more massive than the one that collapsed. "Unfortunately," notes Israel, "it is an idea that cannot be proved —or refuted."

10 HOW MANY GALAXIES ARE IN THE UNIVERSE?

The total number of galaxies in the universe is at least 20 billion and probably well over 100 billion. The average galaxy is a city of 100 billion stars, although much larger as well as smaller galaxies are known.

These are truly colossal numbers, in the same league as our mind-numbing federal deficit. But the federal-deficit number is the result, presumably, of simple addition—somebody added it up. The number of stars in our galaxy was arrived at indirectly.

The Hubble Space Telescope detected this cluster of galaxies approximately four billion light-years from Earth. Each oval object is a galaxy about the size of the Milky Way. Space Telescope Science Institute photograph.

Since each star's velocity in its orbit around the galactic core is gravitationally controlled by the total mass of the galaxy, it is possible to estimate that mass by watching the motions of a relatively small number of stars. The result: The Milky Way Galaxy "weighs" about 200 billion times as much as the sun.

An even simpler shortcut allows an estimate of the total number of galaxies in the universe. All it takes is a state-of-the-art telescope photograph of a small section of the sky. Just count the galaxies, then multiply by the number of photographs it would take to cover the entire sky.

In an area the size of your thumbnail held to the sky at arm's length, the electronic-imaging systems of the world's largest telescopes reveal 50,000 galaxies as they penetrate to magnitude 26—100 million times dimmer than the faintest object visible to the naked eye on a perfectly dark night. When we peer to such vast distances, galaxies look like so many raindrops frozen in flight, although in reality they are millions of light-years apart.

The photograph I selected for an estimate of the number of galaxies in the universe is typical of the new-generation electronic images that are much more sensitive than photographic film. It shows about 2,000 galaxies in an area covering only one ten-millionth of the entire sky. The calculation is simple: 2,000 times 10 million, which gives 20 billion galaxies. Other state-of-the-art deep-space images yield similar results.

The problem with this technique is that an unknown number of galaxies are either too dim or too remote to be picked up. But it does provide us with a minimum number. As telescopes improve in the years to come, the number of visible galaxies will inevitably increase. In the meantime, we know that our Milky Way is one of *at least* 20 billion galaxies in the universe.

11 *WHAT IS THE DARK-MATTER MYSTERY?*

Research over the past two decades has uncovered unequivocal evidence that vast amounts of some phantom material pervade the cosmos. Like the fictional opera-house phantom, who secretly inhabited the building yet persistently eluded detection, this mysterious "dark matter," as astronomers call it, is so effectively concealed from view that it is completely invisible, even though it may make up 90 percent of the mass of the universe.

The phantom matter has never been seen directly, but it can be detected indirectly. Galaxies seem to have massive haloes of invisible stuff that manifests itself when astronomers measure the gravitational effect of one galaxy on the motion of its neighbour. A typical galaxy's gravitational pull is apparently several times larger than can be accounted for once we add up all the galaxy's visible components.

The dark-matter enigma has been around since the 1930s, when Caltech astronomer Fritz Zwicky first noticed that the galaxies in the distant Coma galaxy cluster were moving as if they were being pulled

The elegant spiral galaxy M83, some 15 million light-years distant, is but one of roughly 100 billion galaxies that populate the universe. Individual stars seen in this and most celestial photographs are stars in our own galaxy, typically less than 5,000 light-years away. Harvard College Observatory photograph.

by far more gravitational attraction than could be accounted for by the combined mass of the galaxies in the cluster. Zwicky concluded that something he couldn't see was contributing to the cluster's mass.

Today, evidence of the invisible dark matter is overwhelming. It exists in and around not only our galaxy but just about every other galaxy that has been investigated. Dark matter outweighs normal matter—gas, dust, stars, planets, galaxies—by at least 10 to 1. What this means is that matter we are familiar with is actually less than 10 percent of the universe. It is merely suspended in an ocean of something else—something utterly invisible that reveals itself only by its gravitational attraction.

What is this invisible stuff? Since black holes are massive yet hard to find, they would seem the logical candidates. However, most astronomers are convinced that because there would have to be so many of them, we would have detected them long ago. Other candidates include zillions of massive planets like Jupiter floating between the stars and theoretical subatomic particles called axions and wimps (weakly interacting massive particles).

The fact is, we don't know what the dark matter is. It represents the number-one mystery of the universe in the 1990s. But if phantom matter is 90 percent of what's out there, it will control the destiny of the universe by stopping the universe's expansion, perhaps causing it to collapse on itself tens of billions of years from now.

12 WHICH WAY IS EARTH MOVING THROUGH SPACE, AND HOW FAST IS IT GOING?

Every second of every day, Earth hurtles through space on a never-ending journey. First, the planet is rotating on its axis at a speed of 1,660 kilometres per hour at the equator. At middle latitudes, the velocity is about 1,100 kilometres per hour. Of course, we don't feel this motion, because the Earth's gravity overrides the effect, keeping the atmosphere and other objects in place as the planet spins. But the motion is obvious in another way: it causes day and night as we face alternately toward and away from the sun.

Next, Earth orbits the sun once a year at a speed of 108,000 kilometres per hour. One by-product of the annual orbit is the seasons; another is the march of the seasonal constellations as, in the Earth's trek around the sun, its night side faces different regions of the galaxy. But motions beyond the daily axis rotation and orbital revolution around the sun are imperceptible, even to the careful observer.

Carrying Earth and the rest of the solar system with it, the sun is heading in the direction of the star 99 Herculis at 900,000 kilometres per hour. This motion is the result of the sun's orbit around the Milky Way Galaxy—an orbit so enormous that the sun and its family take about 180,000 years to complete one circuit. During this odyssey around the galaxy, the sun's distance from the galactic nucleus

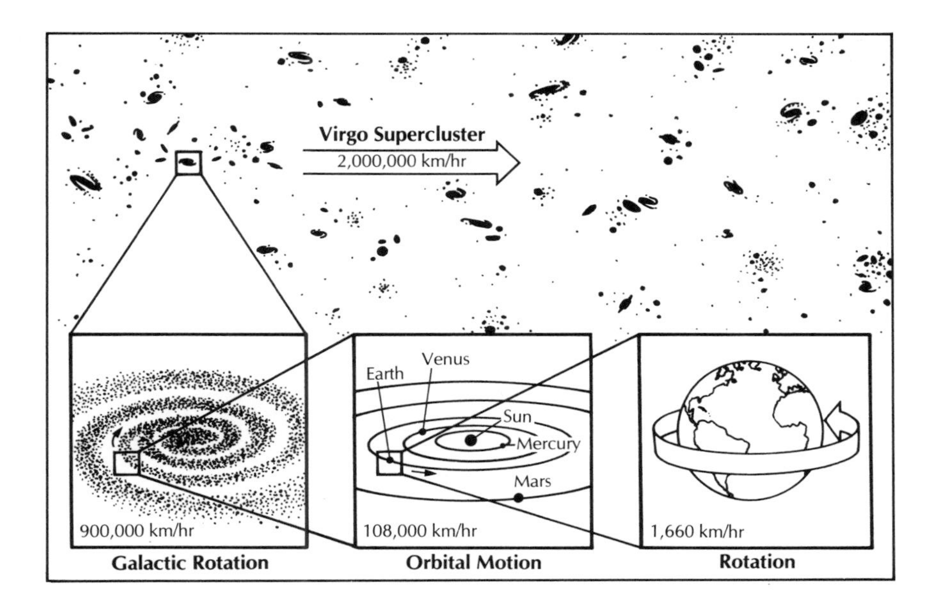

ranges from about 24,000 to 26,000 light-years.

The next tier is the motion of the entire Milky Way Galaxy. Our galaxy lies on the outer fringe of a great continent of at least 10,000 galaxies called the Virgo supercluster, or the local supercluster. The centre of the supercluster is 60 million light-years away, in the same line of sight as the constellation Virgo. Many similar galaxy superclusters are known, but only recently have astronomers detected a pattern in their motion.

The Virgo supercluster, along with several other superclusters, is being pulled toward a region of space that has a somewhat higher-than-average concentration of galaxy clusters. The motion is in the direction of the constellation Centaurus, at a velocity of two million kilometres per hour. In nine months, our galaxy moves the equivalent of the diameter of Pluto's orbit because of this effect.

However, none of this overrides the expansion of the universe, which carries galaxy clusters away from each other more rapidly than the concentration of galaxies is pulling us toward it. But the uniformly expanding universe acts as a frame of reference, and against that, we are shifting at two million kilometres per hour. What it all means is that we are going nowhere fast.

The main motions of the Earth in space are: axis rotation; orbital revolution around the sun; and the sun's galactic path as it moves with the turning of the Milky Way Galaxy. The galaxy is part of the local supercluster of galaxies, called the Virgo supercluster, which is moving in its entirety toward a vast concentration of galaxies in the direction of the constellation Centaurus.

STARS

Every star seen on the darkest, clearest night in the country is part of our galaxy, the Milky Way. Moreover, those stars are the sun's galactic neighbours. All of them are in the same region of the spiral arm of the Milky Way that our sun inhabits. Even the telescopic photographs in this book that seem to be plastered with stars are just magnified views of the nearer parts of our galaxy.

Widening the scene, the Milky Way Galaxy is but one of billions of galaxies in the universe, each with hundreds of billions of stars, though only the highest-resolution telescopic views of nearby galaxies actually reveal individual stars. Stars are the citizens of the universe; galaxies are the cities.

To me, the most striking aspect of astronomical photographs speckled with stars is the fact that every one of those specks is a sun. Many could be accompanied by families of planets. Some of those planets could be Earthlike. A few could harbour life, possibly intelligent life. The question of life beyond Earth

On the darkest nights, far from city lights, the sky seems to brim with stars. In this scene, taken in rural eastern Ontario by the author, the constellation Orion is just above centre. The bright star to the left of centre is Sirius.

is as old as astronomy itself. To know the stars is to be constantly reminded of who we are and where we are.

13 WHAT IS THE BRIGHTEST STAR?

Ask anyone to name the brightest star in the night sky. Chances are, the answer will be the North Star. It is a widespread misconception that the North Star (called Polaris by astronomers) outshines all the other stars. Actually, it ranks thirty-fifth among the stars visible from midnorthern latitudes.

The number-one position is instead held by Sirius, a star that, in general, suffers from a lack of name recognition. Somewhat unglamorously, Sirius means "dog star," because it is the brightest star in the constellation Canis Major, the big dog. You can find Sirius in the south on any clear evening from December through April. It has a bluish white hue and is often twinkling more vigorously than any other star. Only the planets Venus and Jupiter regularly outshine Sirius. But they are planets. Among true stars, Sirius is tops.

I am reminded of the lack of recognition that Sirius receives every time I give a talk to schoolchildren. Rarely have any of them even heard of this star. This does not surprise me, since few adults are familiar with Sirius either, but I was impressed by a question I once fielded from a grade five student. He asked whether Sirius really was the brightest star of all or whether it just looked that way.

The student was right in thinking that appearances can be deceiving. Stars are suns, basically like our own. (Which means, of course, that the sun is actually the brightest star seen from Earth. But here, we are referring to stars of the night sky. On Earth, we see the sun only during the day.) Stars come in a wide range of sizes and brightnesses. Sirius is 23 times as luminous as the sun, but it is also only nine light-years away, nearby in astronomical terms. It is its nearness rather than its actual luminosity that makes it a dazzler in the Earth's night sky.

What is the true king of the galaxy? Among the easily recognized stars, Rigel, the brightest star in the constellation Orion, is 150,000 times brighter than

The galaxy NGC 891, seen edge on, is an 80,000-light-year-wide disc of stars similar to the Milky Way. The irregular black material in the plane of the disc is dust and gas—nebulas from which new stars will eventually be born. Such clouds in our own galaxy are seen as dark patches in the Milky Way. Photograph by Tony Hallas and Daphne Mount.

the sun. It appears dimmer than Sirius because it is more than 100 times farther away. If less conspicuous stars are included in the inventory, even Rigel pales. Although Rigel is the most luminous of the well-known stars, it ranks about tenth in the galaxy.

Among the measured stars, the crown goes to an obscure star known as HD93129A, located 11,000 light-years away in the southern constellation Carina. HD93129A is about one million times brighter than the sun, although it cannot be seen with the unaided eye. The most luminous naked-eye star is Rho Cassiopeiae, which is about 500,000 times as bright as the sun.

14 IF A STAR'S LIGHT TAKES THOUSANDS OF YEARS TO REACH US, ARE THE STARS WE SEE TONIGHT STILL THERE?

The answer to this question is almost certainly yes, for two reasons. Stars have long lives, and they move through space at a relatively leisurely pace.

Even the most distant, reasonably bright star visible to the unaided eye is less than 4,000 light-years away. The nearest star is just over 4 light-years away.

Portraits of the night sky which show so many stars that they seem to be touching each other give the impression that stars are actually close together in space. This is an illusion. One star may be 100 light-years away, while its immediate neighbour might be 1,000 light-years distant. Photograph by Tom Dey.

Depending on their distance from Earth, we see individual stars as they were anywhere from 4 to 4,000 years ago. Because the amount of time it takes the light to reach us is only a tiny fraction of a typical star's life span, you can be virtually certain that any star you look at is still shining in today's time.

Most stars remain basically unchanged for millions or billions of years, merrily radiating light and other forms of radiation at a constant rate. The sun, for instance, will appear essentially as it does now for another five billion years. More massive stars don't last as long, but the odds are good that even short-lived supermassive stars like Deneb and Rigel will appear as they do now in 10,000 A.D.

It is highly unlikely that even one of the few thousand stars seen on a dark night has ended its life in a supernova explosion during the last few millennia. Statistically, it is the same as assuming that in a concert hall of 4,000 people (equivalent to the number of stars visible to the naked eye on a clear, moonless night), no one is likely to die of natural causes during the performance. It could happen, but the odds are heavily stacked against it.

Stars do explode, of course. But so far, all the exploding stars observed have been well below unaided-eye visibility before they erupted. We see the same stars shining in the same places in the sky in which our grandparents saw them—and their grandparents before them. In fact, not one of the 1,000 stars on the first full-sky star map, made by the Greek astronomer Hipparchus, has exploded, disappeared or significantly changed position or brightness since the map was made in 133 B.C.

Although violent action occurs elsewhere in the universe, the starry sky we live under is a serene pasture on the cosmic landscape, far from the hubs of activity.

15 *IS THE SPACE BETWEEN THE STARS REALLY A VACUUM?*

Almost, but not quite. In the solar system—that is, near Earth and the other planets—there is an average of eight atoms of matter—mostly hydrogen—per cubic centimetre. (A cubic centimetre would fill a thimble.) Compare that with the air you are breathing right now, which has 100 million trillion atoms per cubic centimetre.

Between the stars in our galaxy, the density drops to a bit less than one atom per cubic centimetre. Outside the galaxies, in true deep space, there are only a few atoms per cubic metre. While still not a true vacuum, it's darn close.

16 *WHAT IS THE AVERAGE DISTANCE BETWEEN STARS?*

In the vicinity of the sun, the average separation between stars is seven light-years. The sun's nearest neighbour star, Proxima Centauri—a component of the Alpha Centauri triple-star system—is 4.24 light-

Dense spherical swarms of stars called globular clusters, such as the Omega Centauri cluster pictured here, contain several hundred thousand stars within a radius of 50 light-years or less. More than 140 of these clusters orbit the Milky Way Galaxy as satellite star systems. At the core of a globular cluster, stars are separated by light-days, rather than light-years as in our region of the Milky Way. (As a point of comparison, Neptune is 4 light-hours from Earth.) Observations made by the Hubble Space Telescope in 1993 indicate that some stars in globular clusters occasionally venture as close to each other as Uranus and Neptune are to our sun. Photograph courtesy AURA.

years away, or 40 trillion kilometres. That's very generous spacing.

As the stars in the sun's neighbourhood orbit around the hub of the Milky Way Galaxy, they occasionally pass closer to each other. According to calculations by Scott Tremaine, director of the Canadian Institute for Theoretical Astrophysics, a star sweeps to within half a light-year of the sun about once every 30 million to 40 million years. Tremaine estimates the odds are that sometime during the history of the solar system, a star probably came as close to the sun as 0.02 light-years—roughly 40 times the distance from the sun to Neptune.

None of these stellar flybys would disrupt the orbits of the planets, but any star that cruises within half a light-year of Earth would likely disturb the comet reservoir beyond Pluto and cause a comet shower—a diversion of thousands or millions of comets into the region where the planets orbit. Unlike harmless meteor showers, impacts from comet showers could produce mass destruction, like the catastrophe that doomed the dinosaurs 65 million years ago.

Elsewhere in the galaxy, the density of stars varies somewhat. As you would expect, the closer stars are to the galaxy's centre, the more densely packed they are. But the real difference occurs in the galactic-core region. In the 400-cubic-light-year zone at the galaxy's nucleus—a volume in which only a single star would be found in the spiral arms—there are more than two million stars. The average distance between them is just 20 times the diameter of Pluto's orbit, a little over one light-week. For planets of galaxy-core stars, there is no night. The sky would be ablaze with dozens of stars each equal in brightness to that of the moon in our sky.

17 *HOW DO WE KNOW THE DISTANCES TO THE STARS?*

If the distance from Earth to the moon were reduced to the same distance this page is from your eyes, the most distant planets—Neptune and Pluto—would be six kilometres away. But the nearest star beyond the sun, Alpha Centauri, would be 30,000 kilometres away, almost one-tenth the *real* distance to the moon. The stars are very remote. Just as amazing is how astronomers measure such colossal distances.

The first star distance was determined more than a century and a half ago by German astronomer Friedrich Wilhelm Bessel, who used a method that has been employed by astronomers ever since. First, a star's position in the sky relative to the stars that appear around it is recorded. Today, this is done photographically. Bessel, working in the 1830s, did it by eye. Six months later, when Earth has swung to the other side of its orbit, the process is repeated. If the star is within about 200 light-years of Earth, it will have shifted against the background pattern of more distant stars, because we are viewing it at a slightly different angle from the other side of the Earth's orbit. Measurement of the shift, called parallax, gives the distance between the star and Earth.

Parallax measurements produce very accurate distances to stars within 50 light-years of Earth. Between 50 and 100 light-years, this method still gives good results. After that, accuracy declines rapidly, and beyond 200 light-years, educated guesses are just as precise. Six thousand relatively nearby stars have been measured this way, but there are limitations to the technique. Ever present turbulence in the air distorts star images and severely hinders further gains despite the availability of improved tele-

Cepheid variable stars in the Andromeda Galaxy were first discovered in the 1920s when photographs such as this one, showing part of the galaxy's spiral arms, were examined. The technique is still used today, and Cepheid variable stars remain a vital link in distance estimates out to 50 million light-years. Mount Wilson Observatory photograph.

scopes. However, the calculation of star distances is poised for a revolution, a breakthrough as significant as Bessel's first determination.

During the 1990s, two new space observatories will provide a quantum leap in our knowledge of cosmic dimensions. Most productive in this task will be Hipparcos, a European Space Agency satellite designed specifically to measure parallaxes of the brightest 100,000 stars with 10 to 100 times the accuracy that Earth-based telescopes can achieve. Launched in 1988, Hipparcos is still collecting data. When the satellite's readings are fully analyzed later in the decade, a huge list of more accurate star distances will be compiled. Almost every naked-eye star will have its distance updated, which will be both a boon to research astronomers and a welcome gift to stargazers.

The Hubble Space Telescope, the most expensive and the most complex scientific space hardware ever built, is capable of measuring parallaxes out to 1,000 light-years with 50 times the precision of today's methods. But because that is not Hubble's main purpose, only a few hundred key stars are likely to be gauged during the next decade. These few will act as a sort of census, in that they will tell us a lot about the distances to similar-looking stars.

Brightness variations of a Cepheid variable star in a nearby galaxy are plainly evident in this pair of Hubble Space Telescope images. Once the period of variation from one maximum to the next is determined, astronomers can pin down the star's intrinsic brightness. Comparing this with the star's apparent brightness yields an estimate of the distance, providing a crucial yardstick for measuring the universe.

Even taking into account the sophisticated methods made available by Hipparcos and Hubble, parallax becomes less and less reliable once we reach beyond 1,000 light-years. The vast majority of stars in our galaxy, not to mention those in other galaxies, are simply too remote for parallax; indeed, they are part of the background that allows the parallax principle to work. Astronomers, by necessity, must then turn to estimates and educated guesses, rather than direct measurements.

Such estimates are based on the notion that stars with the same temperature and spectrum will likely have the same intrinsic brightness, called absolute magnitude, just as two light bulbs with the same wattage will. Since we know the absolute magnitudes of the nearby stars whose distances *can* be measured, a star which is suspected of having the same absolute magnitude (that is, the same temperature and spectrum) as a nearby star but which *appears* dimmer should be proportionally farther away.

Here's an analogy that will make this concept clearer. Imagine looking at a city's streetlights from a hill. The more distant lights appear dimmer, even though they are probably the same wattage as nearby lamps. Assuming you know the wattage (or absolute magnitude) of the lamps as well as the dis-

Star distances out to about 200 light-years are measured by parallax, a technique that uses the width of the Earth's orbit as a baseline for forming a triangle with a star. When Earth is at one side of its orbit, astronomers photograph the target star against the starry background; they then repeat the procedure six months later at the other side of the Earth's orbit. The apparent shift of the target star against the background stars yields the distance. The larger the shift, the closer the star.

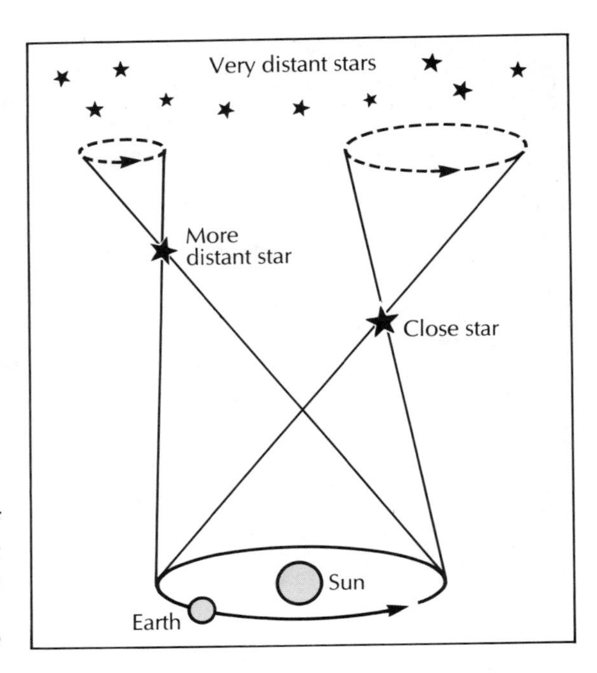

tance to the nearest ones, you could figure out the distances to all the streetlights by measuring their apparent brightness with a sensitive light meter. The light meters astronomers use on their telescopes for measuring star brightness are called photometers.

The next step in the distance hierarchy involves calculating the distances to other galaxies. To gauge these distances, astronomers observe pulsating stars called Cepheid variables. Cepheid variables are crucial distance calibrators, because their intrinsic brightness is known to be related to their period of pulsation. For instance, a Cepheid with a long period of oscillation has high intrinsic brightness, like a 100-watt light bulb, whereas a short-period Cepheid would be a 40-watt light bulb, and so on. Fortunately, Cepheids are typically more than 1,000 times brighter than the sun and can be seen across vast distances. They have been used as distance calibrators out to 30 million light-years, which includes several dozen nearby galaxies.

The distances to more remote galaxies—those situated at the very edge of the visible universe—are estimated from their "redshifts," that is, their velocity of recession from us caused by the expansion of the universe. This method is open to different interpretations because of uncertainties in the rate at which the universe is expanding (for details, see Question 8). Depending on the value assumed for the expansion rate, one team of astronomers might say a galaxy is three billion light-years from us, while another would assign a figure of six billion light-years.

Such wide discrepancies in galaxy distance estimates have been debated for three decades. Astronomers hope that the Hubble Space Telescope and a new generation of giant Earth-based telescopes will soon resolve the discordant distance estimates for galaxies beyond a few million light-years from Earth.

18 WHY DO STARS TWINKLE?

Twinkle, twinkle, little star,
How I wonder what you are.

The short answer to the question posed in this familiar nursery rhyme might be "not a planet." One of the most certain ways we have of making this distinction is that planets virtually never twinkle. The reason for that explains a lot about the difference between planets and stars.

Stars twinkle because they are pinpoint sources of light. In reality, of course, the stars are far from tiny, but their enormous distances from Earth reduce them to dimensions so minuscule that the largest telescopes are unable to reveal them as discs. They are mere point sources. Thus a beam of light from a star entering your eye is a fragile thread that is easily rippled by the ever present turbulence in the Earth's atmosphere, which causes the twinkling.

But a planet is not a pinpoint, it's a definite disc, as even ordinary binoculars will reveal. The bigger bundle of light from the planet is less easily disrupted by atmospheric turbulence.

It's all a matter of *apparent* size. Sirius, for example, one of the nearest stars, is 20 times Jupiter's diameter but 100,000 times farther from Earth. Jupiter therefore appears 5,000 times bigger than Sirius. Compared with the shimmer of Sirius, Jupiter glows with a steady light, unbroken by air currents.

The sky's most obvious twinkler, Sirius is the

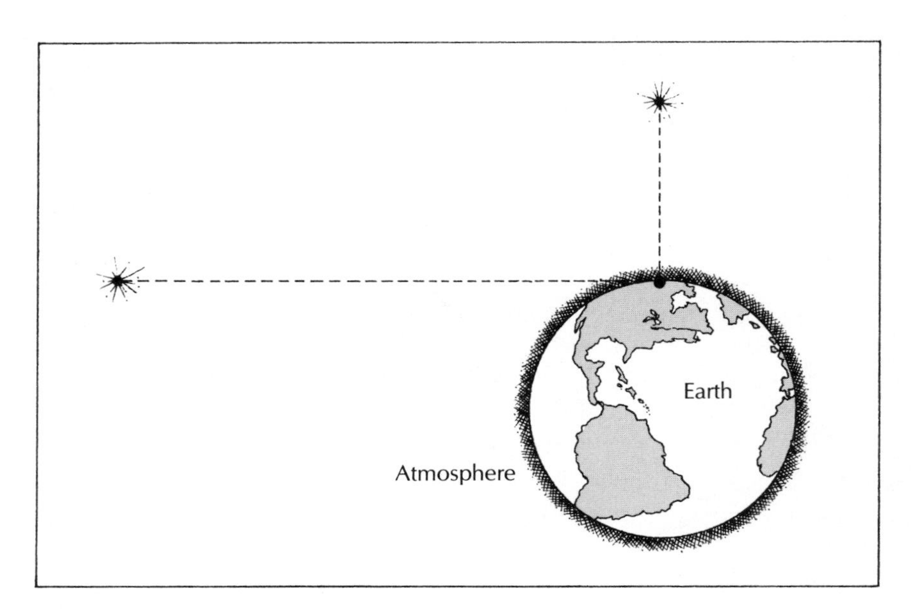

Stars twinkle because their light must pass through the Earth's atmosphere before it reaches the observer. The atmosphere is always turbulent to some degree, which disrupts the steady glow of starlight. The light from stars seen closer to the horizon must travel through more air, and hence those stars are more likely to be obviously twinkling.

brightest object in the southern portion of the evening sky from late December through April. A careful look will show something else: Sirius changes colour. Sirius usually flickers between blue and white, but if the air is especially unsteady, red, yellow and green can be seen interspersed. The phenomenon is best observed when Sirius, or any other bright star, is near the horizon.

These rapid colour shifts are caused by an effect called refraction, wherein the atmosphere acts as a prism, bending different colours of light varying amounts. As turbulent air disrupts the tiny beam of starlight, there are instants when one or another colour predominates during the twinkling.

19 *WILL THE SUN EVER EXPLODE IN A SUPERNOVA?*

The sun will go out with a whimper, not a bang. A supernova—the explosive self-annihilation of a star—is the doomsday fate of stars significantly more massive than the sun.

Theories of stellar evolution predict that the sun will continue radiating essentially as it does now for another five billion years. During that time, it will

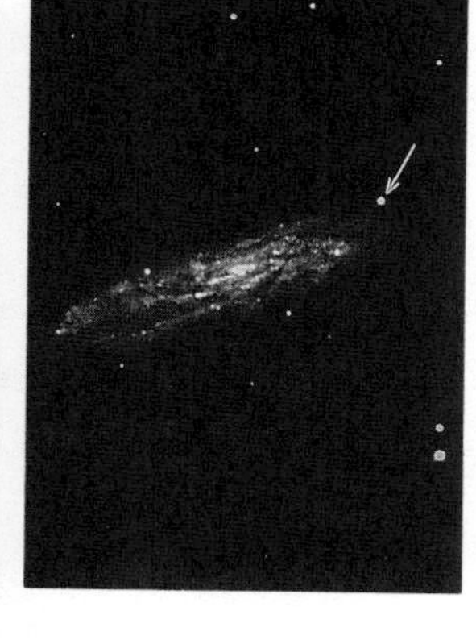

When a star explodes as a supernova, the colossal energy release creates a fireball that may become as bright as an entire galaxy. The arrow in the photograph above points to a supernova at maximum brightness in a distant galaxy. At left are the wispy, gaseous remains of a star that suffered self-annihilation about 30,000 years ago. Supernova remnants mix with existing nebulas, supplying the galaxy with material for future generations of stellar birth. Photographs by Tom Dey (left) and Lick Observatory (above).

gradually elevate its total energy output to about double today's value. Then a profound metamorphosis will begin as the sun switches fuel from "regular" to "super." Over a period of a few million years, hydrogen fusion will end at the sun's core and helium fusion will begin. Because helium fusion is a hotter process with a more vigorous energy output, the pressure from the added energy outflow from the core will bloat the sun to more than 50 times its present diameter. Suddenly, at least in astronomical terms, the sun will become what astronomers call a red giant star.

The sun's red-giant stage will last just a few hundred million years. By the end of it, our star may balloon to more than 100 times its present diameter, consuming the planets Mercury and Venus in the process. Astronomers are uncertain whether Earth will survive the red-giant fires, but if it does, it will be buffeted by intense solar wind as the sun's outer layers are puffed off and its central furnace gradually shuts down.

About six billion years from now, exhausted of fuel after 11 billion years of dazzling radiance, the sun will collapse to a white dwarf star, a sphere approximately the size of Earth but with 70 percent of the sun's mass (the rest will have been lost via the solar wind during and following the red-giant stage). Thousands of white dwarf stars have been observed and catalogued—tombstones that mark the places where luminous stars once shone.

The Earth-sized white dwarf that our sun will evolve into will shine with one three-hundredth the brightness of today's sun (about 500 times brighter than the full moon). White dwarfs get their luminosity from heat that is generated by gravitational compression, rather than from the nuclear fusion which currently powers the sun. These tiny stars are incredibly dense. A teaspoonful of white-dwarf material would weigh one ton on Earth. A lot of energy can be generated by very little squeezing of such dense matter.

The white dwarf will probably outlast the sun's earlier phases. Its estimated life span is 15 billion years, during which it will slowly cool until it is finally a cinder called a black dwarf. No black dwarfs have been detected, and none are expected to exist yet, because the universe is not old enough for them to have evolved.

Stars more than about eight times the sun's mass become even larger red giants, but at the end of the red-giant stage, they undergo a catastrophic collapse, rather than quietly shrinking to a white dwarf. The collapse triggers a colossal explosion that blasts most of the star to smithereens—except for the core, which implodes into either a white dwarf, a neutron star or a black hole. (For neutron-star description, see Question 24; for black holes, see below.)

20 *WHAT IS A BLACK HOLE?*

Black holes reign supreme in our universe. Nothing, not even light, can escape once it falls into the grip of these invisible gravity whirlpools. Celestial vacuum cleaners that can rip apart and devour anything

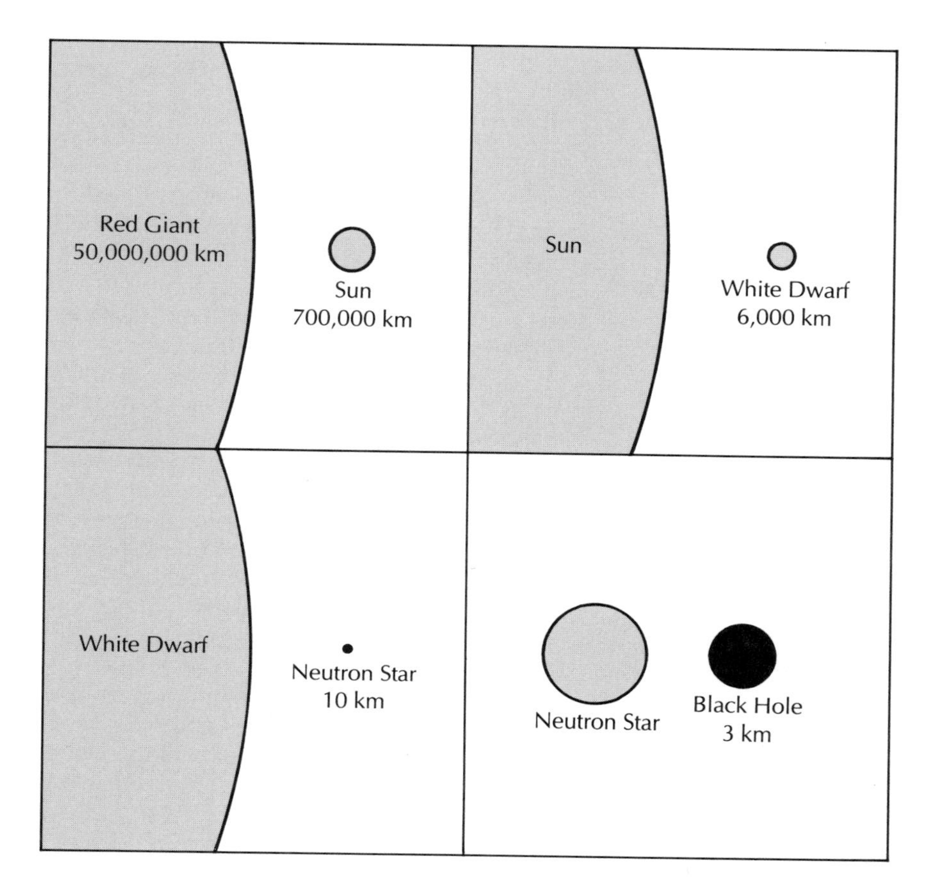

Stars come in a vast range of sizes and densities, and they evolve from one type to another during their lives. Our sun, for instance, will become a red giant in five billion years; then a few hundred million years later, it will collapse to a white dwarf. More massive stars evolve into neutron stars or black holes. (Dimensions given are typical radii.)

from atomic nuclei to entire stars, black holes are almost certainly responsible for quasars, the most prodigious fountains of radiation in the cosmos. Yet little more than a generation ago, black holes were an as-yet-unnamed minor theoretical curiosity.

Astronomers have conceived of such a phenomenon for more than two centuries. In 1784, John Michell, an English cleric, amateur astronomer and the originator of seismology, attempted the first prediction of the ultimate extension of gravity's power. Using Newton's formulas, Michell calculated that in order for a particle to escape into space from the sun's surface, it has to overcome the sun's gravity. Since the velocity of light was known with reasonable accuracy in Michell's time, he was able to calculate that the escaping particle would have to attain one five-hundredth that velocity.

Michell reasoned that if the mass of the sun were somehow increased by a factor of 500, the escape velocity would exceed the speed of light. He concluded that "all light emitted from such a body would be made to return toward it by its own proper gravity." Such an object, he surmised, would be invisible to distant observers. This is the first known prediction of the properties of a black hole.

In 1916, Einstein's colleague Karl Schwarzschild made the first calculations showing that if a body the same mass as the sun could be compressed to an object only six kilometres across, it would dig itself a black hole in space and disappear into it, leaving behind nothing except its gravitational field. This critical size, called the event horizon, marks the black hole's boundary, the point of no return for light or anything else that ventures this close. The event horizon's size varies in direct proportion to the mass of the object that collapsed to create the hole. A black hole with the Earth's mass would have an event horizon the size of a hazelnut.

But neither Earth nor the sun can become a black hole. They aren't massive enough for gravity to have the last word. Earth is as compressed as it will ever get, and the sun will never shrink past the white-dwarf stage, in which atomic nuclei swim in a dense sea of electrons.

Only stars much more massive than the sun can crush themselves into the final abyss of a black hole. It happens in the fireball of a supernova, an explosive stellar annihilation that is triggered by the star's sudden collapse when it runs out of thermonuclear fuel around its core. During the blast, the star's core implodes. If the core exceeds four times the mass of the sun (as it probably would if the original star were more than 20 times the sun's mass), nothing can stop the collapse, and a black hole is born.

21

IF BLACK HOLES ARE INVISIBLE, HOW DO WE KNOW THEY EXIST?

As recently as the late 1960s, black holes were mathematical curiosities rarely discussed in scientific literature. They didn't even have a name until physicist John Wheeler coined the term "black hole"

The black hole Cygnus X-1 is at the centre of the gravitational whirlpool in orbit around a giant blue star. Illustration by Adolf Schaller.

in a 1968 article in *American Scientist* magazine. Around the same time, theorists began thinking about how a black hole might be detected. They suggested that matter swirling into a black hole would be heated to billions of degrees and would radiate bursts of x-rays before plunging across the hole's event horizon.

In 1971, the first x-ray satellite, called Uhuru, detected a celestial source of radiation known as Cygnus X-1 that appeared to be behaving in a manner suspiciously like the predictions for a black hole. The Uhuru positional information about Cygnus X-1 led astronomers using ground-based telescopes to a seemingly ordinary blue giant star in the same spot in the constellation Cygnus. Using the University of Toronto 74-inch reflector telescope in Richmond Hill, Ontario, astronomer Tom Bolton obtained spectrograms which showed that the blue giant, about 27 times the mass of the sun, orbits the x-ray source once every 5.6 days. The orbital data mean that the two objects are separated by less than three times the blue giant's diameter.

More important, Bolton's subsequent research strongly indicates that Cygnus X-1 must be at least six times the sun's mass and is most likely 15 times more massive than the sun—well above the four-solar-mass theoretical minimum for black holes. The only reasonable scenario is that the x-rays are being emitted by material from the blue star swirling around and falling into a black hole. At one time, the hole must have been a star, most likely a giant similar to its present companion.

If Cygnus X-1 were the only example of a black hole, astronomers would still be skeptical about the reality of the exotic beasts. But there are at least four other systems in which compact x-ray objects and ordinary stars are locked in a mortal embrace, just like Cygnus X-1 and its blue-giant partner. A very solid case has been made indicating that at least a few black holes have been found.

Furthermore, there is strong circumstantial evidence that a black hole several million times the sun's mass marks the centre of our Milky Way Galaxy. Many other galaxies also seem to harbour black holes at their cores, some of which are more than a billion times the sun's mass. The existence of these monster galactic black holes was deduced by measuring the velocities of stars and gas whirling around the cores of the galaxies.

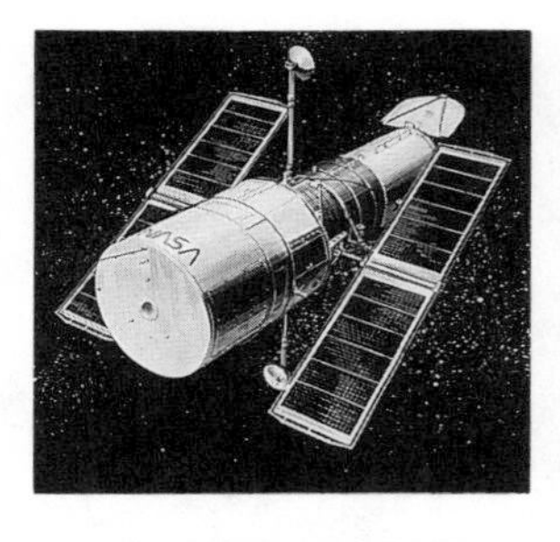

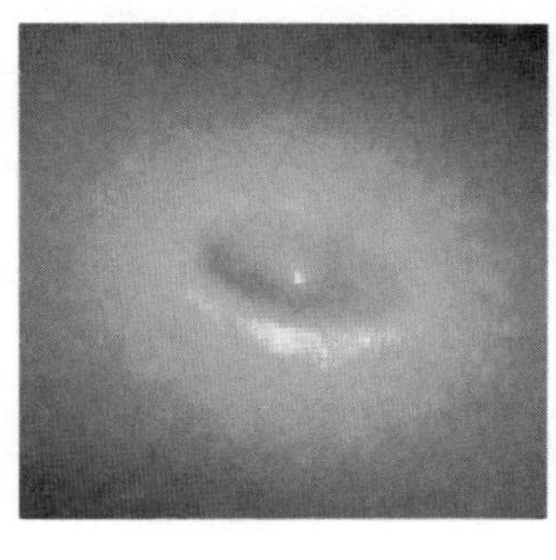

This image of a suspected black hole at the core of the galaxy NGC 4261, about 50 million light-years distant, was obtained in 1992 by the Hubble Space Telescope. The hole itself is invisible, buried in the bright region at the centre, but the doughnut around it is a 500-light-year-wide accretion disc—a giant celestial whirlpool of gas and dust being funnelled into the hole. Space Telescope Science Institute photograph.

22 CAN A BLACK HOLE BE USED AS A TUNNEL THROUGH SPACE?

In 1971, there was a flurry of excitement when the first black hole was identified after radiation emitted by material flowing into the hole had been observed. That got theoreticians thinking about the immense warping of space and time that must occur around and inside these bizarre objects and how they might be utilized as time tunnels by space explorers sometime in the remote future. But it soon became clear that any spacecraft plunging into the space-warping regions of a black hole would be ripped atom from atom on the way through. The idea of black holes as hyperspace gateways soon fizzled out.

In 1985, astronomer Carl Sagan was working on his science fiction novel *Contact*. The plot called for an alien civilization to travel vast distances faster than the speed of light. Rather than resort to a fictional shortcut through some vaguely described hyperspace, as hundreds of science fiction writers had done before, Sagan telephoned Caltech gravitation theorist Kip Thorne and asked him whether he could think of any scientifically plausible way that it could be accomplished. Thorne, along with graduate students Michael Morris and Ulvi Yurtsever, quickly immersed himself in the project.

The team soon confirmed that the atom-crushing forces within and near black holes could never conceivably be beaten back to accommodate human travel. Instead, they found that a cosmic "wormhole" —a purely theoretical structure that may or may not exist in nature—could be used as a traversable tunnel through space and time. A wormhole, according to Thorne, is a tubular distortion in space that "links widely separated regions of the universe."

Wormholes do not defy the general theory of relativity, yet they allow virtually instantaneous travel across any distance, even billions of light-years. Hypothetically, one could travel into both the future and the past *anywhere* in the universe.

"If the laws of physics permit traversable wormholes," notes Thorne, "then they probably also permit such a wormhole to be transformed into a time machine." A time traveller would enter the wormhole in his or her "present" and instantly emerge out the other end somewhere in the past or the future. The catch lies in building a protective structure from superdense "exotic" matter that would act like antigravity to keep the wormhole from collapsing. Thorne says this may prove to be an insurmountable problem, but he suggests that the lure of time travel will be so impossible to resist that advanced civilizations will be compelled to delve into the project.

Wormholes have never been detected, and no one is really sure what one would look like, except that only the ends would be visible and that they would superficially resemble black holes. For now, they merely exist in theorists' minds. But consider this: More than five centuries ago, the basic principles of powered flight were worked out by Leonardo da Vinci. Airplane travel had to wait for technology to catch up. Hyperspace and space-warp travel are dreams which will never go away, especially now that they seem possible—in theory.

PLANETS

Planet gallery, clockwise from top: Mercury, Venus, Earth, moon, Mars, Jupiter, Saturn, Uranus, Neptune. As recently as 1970, none of these images existed except the Earth-based photographs of the moon and Mars. Jet Propulsion Laboratory photograph.

Astronomer Carl Sagan once referred to the two decades from 1970 to 1990 as "the Golden Age of Planetary Discovery." He was right. During that period, robot spacecraft landed on Venus and Mars and obtained close-up photographs of all the other planets except Pluto. In 20 years, we learned more about our neighbour planets and their five dozen moons than all the accumulated knowledge gained from centuries of observation from Earth.

The robotic emissaries from Earth discovered volcanoes on Mars that dwarf anything on our planet; geysers spewing methane and liquid nitrogen on Triton, a moon of Neptune; and a subsurface ocean of water on Jupiter's moon Europa. Other finds included the thick nitrogen atmosphere of Saturn's satellite Titan; sinuous erosion channels on Mars, apparently caused by water that once flowed there; and swirling storm cells in the multicoloured atmosphere of Jupiter.

In just one generation, the planets ceased to be re-

mote and largely mysterious shimmering dots in astronomers' telescopes. Today, they are mapped, and their compositions are understood in detail. Children now grow up learning about real places beyond Earth, not just names.

23 DOES PLANET X EXIST?

Whenever I speak about stargazing and astronomy to schoolchildren, I am almost always asked, "What about the tenth planet—Planet X?" Kids love Planet X. I put it down to the enticing mystery of the unknown. The name oozes mystery all right, but does Planet X exist?

The concept of a tenth planet emerged more than 50 years ago when astronomers noticed that the planets Uranus and Neptune deviated slightly from their predicted positions as they progressed along their orbits. At first, observers believed that the gravitational effects of the ninth planet, Pluto, might explain the discrepancies. But, as it turns out, Pluto is far too small to account for them. Instead, the gravitational influence of a tenth planet several times larger than Earth, dubbed Planet X, was offered as an explanation for the observed deviations.

Yet search after search has failed to reveal the hypothetical planet. The most recent effort involved detailed examination of data from a survey by IRAS, the Infrared Astronomical Satellite. IRAS made two scans of 70 percent of the sky in 1983, but it has taken nearly a decade to analyze the data thoroughly. Researchers were intent on finding something that moved during the interval between scans, the way a remote planet would. Ideally suited to tracking down a distant planet, IRAS's telescope nevertheless turned up nothing.

How do we explain the deviations in the orbital positions of Uranus and Neptune? Uranus takes 84 years to orbit the sun, while Neptune requires 165. Since its discovery in 1846, Neptune has yet to complete one orbit of the sun. Part of the problem is that Neptune's orbit is not known with high enough precision to predict the existence of a planet well beyond Pluto. Thus the foundation for theorizing that Planet X exists is almost entirely based on histori-

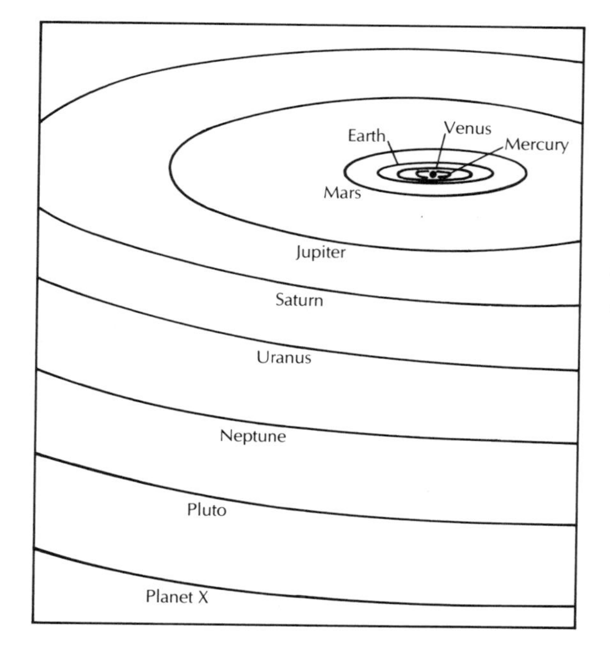

For more than half a century, Planet X was thought to be lurking out beyond the orbit of Pluto. But a 1993 study has finally proved that there is no evidence to support the existence of this hypothetical planet.

cal observations of Uranus's position in its orbit.

In 1993, planetary scientist Myles Standish of the Jet Propulsion Laboratory, in Pasadena, California, produced what appears to be the definitive study on this subject. He found that most of Uranus's apparent deviations could be traced to errors of measurement made at a single observatory between 1895 and 1905. Standish's research showed that the remaining deviations were caused by errors introduced by using the incorrect value for the mass of Neptune when calculating its gravitational influence on Uranus's position. When a more precise value for Neptune's mass was used (determined by the Voyager 2 spacecraft when it passed near Neptune in 1989), the remaining Uranus deviations disappeared. Planet X is now history.

In 1992, just as astronomers were writing off Planet X, something was discovered beyond Pluto. Using an 88-inch telescope atop Hawaii's Mauna Kea, astronomers David Jewitt and Jane Luu spotted an object about 200 kilometres in diameter that appears to be slightly farther from the sun than Pluto. Much too small to be classed as a planet, the as-yet-unnamed object is probably one of the larger members of a belt of comets long thought to exist beyond the orbits of Neptune and Pluto.

24 HAVE ANY PLANETS BEEN SEEN ORBITING OTHER STARS?

There was a time when simply talking about the existence of planets of other stars was dangerous to one's health. In 1600, Giordano Bruno was burned at the stake, at least in part for holding the heretical notion that "innumerable suns [exist], and an infinite number of Earths revolve around them." Today, the search for planets orbiting other stars is one of the great scientific quests of the late 20th century.

Although astronomers suspect that planets like those in our solar system will eventually be discovered orbiting a star similar to our sun, nothing resembling our own situation has as yet been uncovered. And even if duplicates of all nine planets in our solar system were orbiting Alpha Centauri (the closest star beyond the sun), current equipment is nowhere near capable of detecting any of them.

Planets are lost in the glare of their parent stars. Even Jupiter, the largest planet in our solar system, is utterly dwarfed by the sun, which not only is 10 times Jupiter's diameter but glows brilliantly from its own radiation. Jupiter—and every other planet—shines instead by reflected sunlight. Seen from afar, the sun would be one billion times brighter than Jupiter and about 25 billion times brighter than Earth. No wonder direct visual searches for planets of other stars have been unsuccessful.

Not to be denied, however, astronomers have resorted to several indirect techniques, all of which rely on observations of stars, rather than of suspected planets. In one scenario, astronomers search for certain subtle motions of a star that might indicate the gravitational influence of one or more planets. In theory, that should be easier to detect than the planets themselves. In practice, it has turned out to be exceedingly difficult.

The first report of indirect detection of another solar system hit the astronomy world in 1963, when Peter Van de Kamp of Swarthmore College Observatory, in Pennsylvania, claimed that he had found

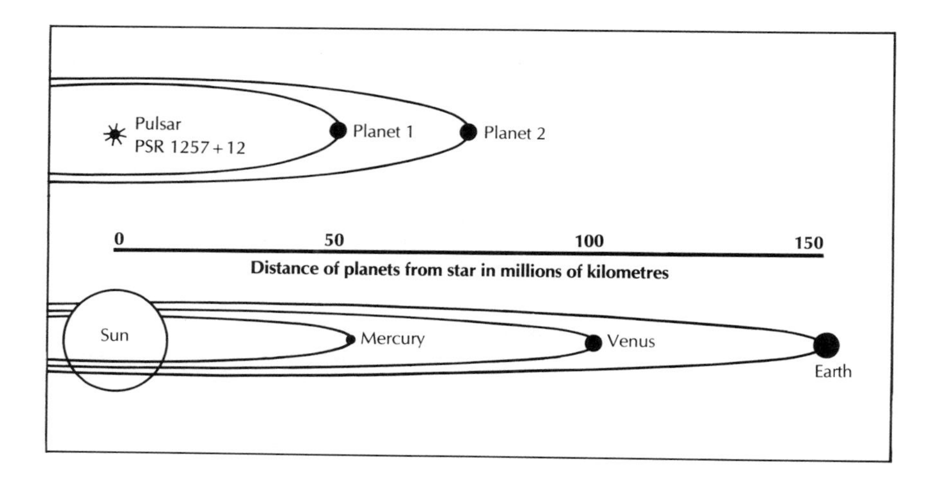

a Jupiter-sized object orbiting Barnard's star, the second nearest star to our solar system. Although Van de Kamp had not actually seen the object, the deviations in the position of Barnard's star, which he believed to be caused by the gravitational tugging of a large planet, were presented as evidence. Six years later, Van de Kamp offered refined measurements that indicated there were not one but two planets like Jupiter orbiting Barnard's star. Further study proved, however, that the deviations were in the Swarthmore telescope, rather than in Barnard's star. And that was just the first such false alarm.

The bizarre PSR1257+12 system is the only known example of a star other than the sun suspected of having an Earth-sized planet in orbit around it. For details, see Question 24.

In my research files, I have a folder titled "Extrasolar Planetary Systems." It bulges with journal articles and magazine and newspaper clippings. Since 1963, there have been at least 13 "discoveries" of planets of other stars. Like the Van de Kamp find, they have withered under the test of time. Some were instrument errors. Others proved to be objects more than 80 times Jupiter's mass—far too massive to be classed as planets (see definition in the Introduction). A few suspects remain at the threshold of detection with modern equipment and cannot be confirmed. Only one case is currently accepted by most astronomers, and it is a strange one that emerged in an entirely unexpected way.

During routine studies of pulsars in 1991, sensitive radio-telescope observations showed tiny variations in the pulses from one of them, PSR1257+12. A pulsar is a rapidly spinning neutron star that radiates energy in pulses as it spins, like a lighthouse beam. Neutron stars are tiny objects by astronomical standards—they are about the same width as a small city but are incredibly dense. PSR1257+12 has a mass equal to 500,000 Earths and spins on its axis 161 times per second. If brought to Earth, a teaspoonful of its material would weigh more than all the concrete, bricks and steel used in every building, bridge and roadway in Canada.

Analysis of the oscillations of PSR1257+12 suggests that the pulsar is being slightly nudged toward and away from us by the gravitational pull of two planets, 3.4 and 2.8 times the Earth's mass, which orbit the pulsar once every 67 and 98 days, respectively. Apparently, the two planets have circular orbits similar in dimensions to the orbits of Mercury and Venus in our own solar system.

Since this was the first time anyone had claimed detection of planets similar in size to Earth, the announcement, made in January 1992, caused quite a sensation. But similarity in size does not mean similarity in nature or origin, as the discoverers—astronomers Dale Frail and Alexander Wolszczan—were careful to point out. Indeed, a pulsar was the last place anyone expected to find a planet. According to theory, neutron stars, which can be observed as pulsars, are the imploded cores of stars that have exploded as supernovas. Theory also suggests that any planets orbiting a star that becomes a supernova would be incinerated or flung into interstellar space or into unstable, elongated orbits.

Frail and Wolszczan believe that the planets they found formed after the supernova explosion which created the pulsar. "The pulsar must have had a companion star similar to the sun," says Frail. "We know this because the pulsar's spin rate has been accelerated. The only way that can happen is if material from the companion cascaded onto the pulsar, speeding it up."

The companion star was ultimately consumed by the pulsar, explains Frail, but in the process, a portion of the star's material would have formed a disc around the pulsar. It is from material in this disc that the pulsar's planets must have emerged. A similar disc is thought to have surrounded our sun 4.6 billion years ago, when our solar system was born.

As bizarre as anything in science fiction, these new planets are 1,000 times larger than the star they orbit. Their sun, the pulsar, would be so dim, it would look like a typical star in their sky. The planets themselves are most likely rocky bodies, both about 1.4 times the Earth's diameter, enriched with iron, gold and other heavy metals. But future space explorers had better beware: Apart from their crushing gravity—roughly twice that which exists on Earth—the two planets are bathed in deadly gamma rays and other lethal particles booming out in prodigious quantities from the pulsar.

Since the consensus among astronomers is that the pulsar planets formed from debris left over from the supernova which created the pulsar, this case does not stand as an example of a solar system like our own. What we want to find are planets orbiting a sunlike star. Nevertheless, pulsar PSR1257+12 has provided the strongest evidence so far for the existence of planets beyond the solar system, regardless of their origin.

One additional piece of evidence in this puzzle was added by the Hubble Space Telescope in 1992. Peering into the Orion Nebula, 1,500 light-years away, the telescope recorded detail never seen before, including 15 infant stars embedded in discs of gas and dust that resemble the theoretical models of the disc from which our solar system formed 4.6 billion years ago. According to the Hubble researchers, the discs typically contain at least 15 times Jupiter's mass in gas and dust, as had been theorized. The find provides compelling photographic evidence of solar systems in formation, which in turn strongly suggests that planetary systems are abundant in the galaxy. It seems only a matter of time before a planetary family similar to our own is discovered.

25

WHAT IS THE "FACE" ON MARS?

In July 1976, a pair of American Viking Orbiter spacecraft began circling Mars and photographing the planet's surface. One of the pictures taken later that year shows a two-kilometre-wide mesa (a roughly table-shaped land formation) which resem-

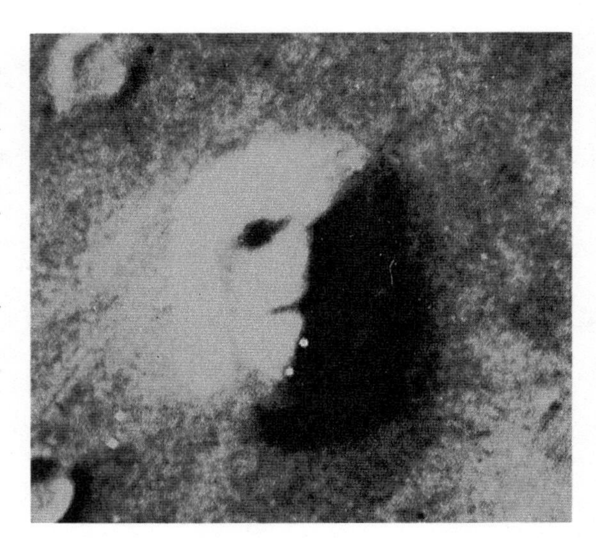

Because the "face" on Mars bears a remarkable resemblance to Homo sapiens, *supermarket tabloids and certain television programmes have cited it as "proof" that alien life exists. However, this feature is almost certainly nothing more than an unusually shaped landform on the red planet. NASA photograph.*

bles a human face staring straight up, like a colossal carving on the arid rocky surface of Mars.

The "face" picture, taken of a region known as Cydonia, was among hundreds released by NASA to the news media as the Viking mission progressed. The feature was explained as "a Martian mesa of unusual shape resembling a face" and was forgotten — for a time. Then it began to appear in publications such as the *National Enquirer*, accompanied by headlines like "Spacecraft Photographs Life on Mars." Today, the face may be the best-known image of the red planet. And the interest goes beyond supermarket tabloids.

In 1979, two computer-image-processing experts, Gregory Molenaar and Vincent DiPietro, decided to investigate the face and found that one of the Viking Orbiters had taken a second picture of the feature at a different sun angle. This image convinced them that the rock mesa bears a striking resemblance to a humanoid face. "If this object was the result of natural forces, it indicates nature is a highly intelligent force," they later wrote in their privately published report.

Next on the scene was science writer Richard C. Hoagland, the spokesperson for Molenaar, DiPietro and a small group of engineers and computer experts interested in analyzing the face images. At a news conference shrewdly timed to coincide with a 1988 Soviet Union spacecraft encounter with Mars, Hoagland said computer enhancement of the pictures added weight to the view that extraterrestrials had constructed the face on our neighbour world as interplanetary graffiti to attract our attention. Hoagland offered what he called compelling evidence that the face and some pyramid-shaped hills nearby were the work of extraterrestrials.

I try to avoid knee-jerk reactions to such sensational claims, so I examined the best available computer-enhanced versions of the photographs myself. I am not at all persuaded that we are seeing the work of alien life. It is much easier for me to believe that of the millions of landforms visible in the Viking photographs, some would inevitably trigger human response, just as a search of pebbles at the lakeshore, if conducted long enough, can reveal a familiar shape. For one thing, the so-called pyramids on Mars are, upon close inspection, inexact, natural-looking features. I have seen similar formations on the moon through my telescope. Moreover, if aliens are pyramid builders, why would they erect crooked structures? A perfect pyramid of the huge dimensions involved would stay virtually unchanged for millions of years on the arid plains of Mars. If, as Hoagland maintains, the face is meant to resemble our evolutionary ancestor, *Homo erectus*, it and the pyramids would be less than two million years old.

We may not have long to wait for the verdict. The U.S. Mars Observer space probe has just reached the planet and gone into orbit around it as I write this (in 1993) and will soon be mapping the entire planet in more detail than ever before. If mission controllers manage to photograph the face in high resolution, I am certain the pictures will show the feature for what it is: a natural rock formation which does indeed resemble a face but which does so *only* under certain conditions of illumination.

26 *IS JUPITER REALLY A FAILED STAR?*

In his television series, *Cosmos*, Carl Sagan calls Jupiter "a star that failed." He makes similar statements in at least two of his popular books.* Sagan may not have been the first to make this claim, but he is certainly responsible for popularizing the notion. I regard it as an unfortunate "sound bite" implying that if Jupiter had been only a bit more massive, it would be a star, rather than a planet.

The reality is that Jupiter is about as far from being a star as Earth is from being a planet like Jupiter. The giant planet would have to be 80 times more massive to reach the minimum threshold mass for a star. Conversely, if Jupiter were 300 times less massive, it would be a planet like Earth. We could say almost as easily that Earth is a failed Jupiter as that Jupiter is a failed star.

One thing Jupiter does have in common with stars is its almost entirely gaseous composition. A colossal globe of hydrogen with generous helpings of helium, methane and ammonia mixed in, Jupiter is a world without a solid surface as we know it. The visible face of Jupiter is cloud tops, a colourful veil of haze and ice crystals.

Jupiter's deep interior will remain off limits to any imaginable exploratory device of the future because the giant planet's enormous bulk creates a crushing internal pressure (simply from the weight of overlying material) that generates a temperature of 30,000 degrees C at the planet's core. Although this might seem starlike, it is far short of the 30,000,000 degrees needed to ignite the thermonuclear reactions that power the stars. By the time Jupiter's central heat percolates its way to the planet's cloud tops, there is not much left. The temperature there is a chilly minus 140 degrees.

But let's suppose Jupiter were 10 times its present mass. The internal temperature would then be greater than 1,000,000 degrees C, and the planet's surface would be well above the boiling point of water. Yet this hypothetical planet would be only slightly larger than Jupiter is right now. Theorists have shown that if more massive planets exist, they would all be approximately the same size. The additional mass would simply be compressed in the core region. Jupiter, regardless of its mass, is about as big as a planet can be.

Hypothetical planets in this transitional mass range—between Jupiter and the smallest-known star—have been named brown dwarfs, although not a single example of a brown dwarf has been uncov-

Jupiter, the solar system's largest planet, is a colossal globe 11 times the Earth's diameter. Topped by a quilt of colourful clouds, Jupiter's thick atmosphere of hydrogen and helium superficially mimics the composition of a star. A solid, rocky core about the size of Earth may exist at the planet's centre. Jet Propulsion Laboratory photograph from Voyager 1.

ered with certainty despite many searches around nearby stars. Astronomers are beginning to suspect that there might be some unknown aspect of star and planet formation which acts against the creation of objects in the substellar, super-Jovian size range. Maybe true failed stars are a cosmic oddity.

**One of these,* Cosmos *(Random House; 1980)—the companion volume to the television series of the same name—is among the best popular-level astronomy books of the last 25 years. Despite my nit-picking here, I am a great admirer of Sagan's work.*

27 DID AN EXPLODING PLANET CAUSE THE ASTEROID BELT?

Originally suggested in 1823 by German astronomer Heinrich Olbers, the exploding-planet theory was popular until the 1960s. Now, it is considered an unlikely scenario. I'll get to the current thinking on the subject shortly. First, a few words about the asteroid belt.

Between the orbits of Mars and Jupiter is a region where millions of mini-planets—the asteroids—roam around the sun in orbits ranging from nearly circular to highly elliptical (oval). The orbits of roughly 5,000 asteroids have been charted with

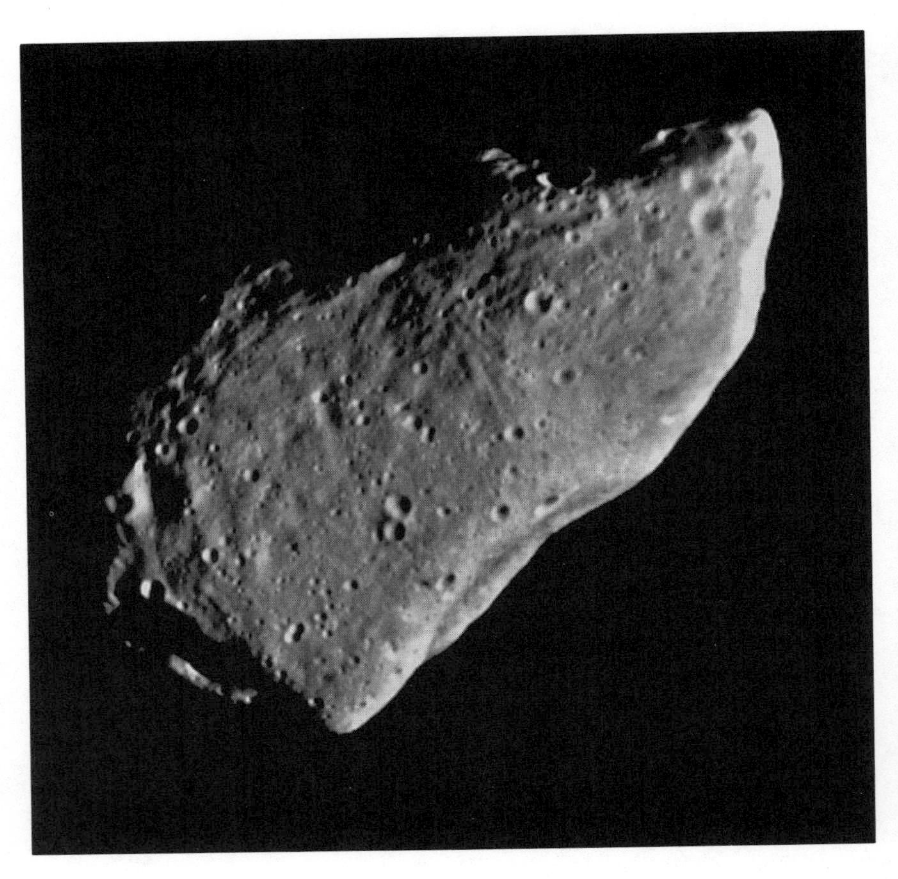

The first close-up photograph of an asteroid was snapped in 1991 by the Galileo space probe during its trek to Jupiter. Named Gaspra, the potato-shaped asteroid is 19 kilometres long and 12 kilometres wide. The smallest visible craterlets are about the size of a football field. Jet Propulsion Laboratory photograph.

reasonable accuracy, and about 200 of them stray outside the main belt zone. A few dozen swing inside the orbit of Earth, while a couple of others venture as far out as Saturn.

The largest asteroid, Ceres, is 915 kilometres in diameter, about the width of France. Ceres orbits the sun once every 4.6 years at 2.8 times the Earth's distance. Some of the other large asteroids are Pallas, 522 kilometres; Vesta, 500 kilometres; and Hygiea, 430 kilometres. Twenty-six asteroids are more than 200 kilometres in diameter, and there are probably at least 100,000 that are more than one kilometre in diameter, millions that are bigger than a barn and billions that are larger than a refrigerator.

The mental image of all that debris roaming the asteroid belt has contributed to its reputation as a dangerous place, a notion which has been fostered in numerous science fiction films. In the second *Star Wars* movie, *The Empire Strikes Back*, the only escape route for the good guys is through the dreaded asteroid field. The intrepid Han Solo plunges his spaceship into the asteroids, jockeying it like a car in a wreck 'em race.

This impression of asteroid belts as outer-space pinball machines, where craggy, crater-pitted boulders are in constant risk of bumping into each other, however, is a total departure from reality. In fact, there is plenty of room between Mars and Jupiter for trillions more asteroids than are already there.

If we were to embark on an interplanetary trip from Earth to Jupiter, cruising right through the asteroid belt, we would probably see fewer than a dozen asteroids during the entire voyage, and those would appear like moderately bright stars, rather than looming boulders.

Current thinking on the origin of the asteroids suggests that they are actually the makings of a planet which never formed, instead of a body that existed and exploded or was smashed. In this scenario, Jupiter has been fingered as the culprit. The disrupting influence of the giant planet's gravity may have permitted many small Ceres-sized subplanets to form but prevented them from coalescing into a larger body, say, the size of Earth. These smaller objects, in turn, collided with one another, creating the multitude of asteroids that exist today. While Ceres and perhaps one or two other large asteroids survived this era intact, most of the smaller bodies must have, over time, been sent on highly elliptical trajectories that eventually resulted in either collision with one of the major planets or ejection from the solar system. Today, the total mass of the asteroids is less than that of our moon.

28 WHAT ARE SATURN'S RINGS MADE OF?

Saturn is bedecked with a hauntingly beautiful ornamentation that has to rate as one of the universe's most dazzling creations. The rings are composed of swarms of icy particles—primarily water ice—ranging from tiny crystals like those in an ice fog to chunks the size of small mountains. Each particle

has its own individual orbit about Saturn, although a gentle jostling occurs as the particles are affected by the gravitational pull of Saturn's major moons.

The rings are truly enormous in extent. From edge to edge, they span a distance equivalent to two-thirds of the gulf between Earth and the moon. Yet the particles that make up the rings are confined to a flat disc no thicker than the height of a 50-storey office tower. A scale model of the rings made of a single sheet of paper the thickness of a page in this book would be larger than a football field.

Because of the distribution of mass within Saturn itself, the rings are not spread in a random haze around the planet. Saturn's 10.7-hour rotation period has produced a bulging waistline—the planet is 120,000 kilometres wide at its equator but only 109,000 kilometres at its poles. Therefore, as a body orbiting around Saturn passes over the equatorial zone, it "feels" a greater gravitational pull than it would over the polar regions, because there is more material below it. The path of greatest orbital stability is a nearly circular one above the most massive sector of the planet, precisely at its equator, and that is where all the ring material gathered long ago.

An exploration of the rings by a spacesuited astronaut outfitted with a propulsion backpack for manoeuvring would be one of the most exquisitely beautiful space excursions that humans could take. By orbiting Saturn in an approximately circular path identical to that of the ring particles, a spacecraft could bring an astronaut safely within a kilometre of the ring structure. A collision with any ring material under these conditions would amount to nothing more than a gentle nudge.

For every house-sized ring boulder, there are a million the size of a baseball and trillions the size of a grain of sand. In denser sections of the rings, the baseball-sized particles are likely separated by just a few metres, while the relatively rare house-sized ones would be kilometres apart. Recent research strongly suggests that Saturn's rings are a temporary structure. Saturn was not born with this adornment. The rings probably formed from debris left over after a comet smashed into one of the planet's satellites. An alternate theory is that two of Saturn's smaller moons collided, and the resulting pieces

Although Saturn's rings look solid from afar, they are actually composed of trillions of individual moonlets, each in its own orbit around the huge gas planet. Jet Propulsion Laboratory photograph (top); illustration by Ludek Pesek (bottom).

subsequently smashed into each other until the relative equilibrium we see today was reached.

Much thinner, less massive rings girding the three other giant planets—Jupiter, Uranus and Neptune—were discovered between 1977 and 1989.

29 *WHY DO THE PLANETS ALL ORBIT THE SUN IN THE SAME PLANE?*

I often tell my astronomy students that as Earth and the other planets in our solar system move around the sun, their paths are confined to a plane, like the balls on the surface of a billiard table. It is a simple analogy, but it works, because one can immediately grasp why the planets (and the moon as well) are always seen restricted to a specific zone—astronomers call it the ecliptic, and we know it as the zodiac—that encircles the sky. A related fact of interest is that all nine planets orbit the sun in the same direction (counterclockwise as seen from above the Earth's North Pole). But the real question here is, Why *do* the planets orbit the sun in a flat plane instead of randomly angled paths?

The pancake orientation seems to date from the solar system's formation 4.6 billion years ago, when the sun and the planets were born from a cloud of dust and gas called the solar nebula. At the nebula's centre, the primordial sun's gravity had corralled at least half of the available matter. The remainder surrounded the sun in a flat disc, like an old vinyl long-playing record, with the sun as the central hole. Within this disc, the planets formed. When the sun's interior reached a temperature and pressure great enough to fuse hydrogen atoms into helium, our star was born. Its radiant energy soon swept away the remnants of the solar nebula, leaving the planets in orbits probably similar to those they travel today.

Astrophysicists who have developed computer models of contracting irregularly shaped nebulas like the one that might have formed the solar system say that the flat-disc arrangement is a natural development. The simulations show that the cloud never collapses uniformly. There is always a preferential motion among the myriad particles that ultimately induces rotation. However slight it may be at first,

All illustrations of the solar system, including this one from NASA showing the flight path of one of the Voyager space probes, depict the orbits of the planets in a flat plane. In reality, all the planets except Pluto do, in fact, move around the sun as if on a flat surface. For the reason why, see Question 29

the rotation of the primordial solar system nebula soon develops into a significant force that twirls the contracting cloud into a colossal pizza shape. The sequence is controlled by the same principle that accelerates an ice skater's spin when the arms are pulled toward the body.

The zodiac is, in effect, the solar system's equator. It has a north pole too. While the star Polaris happens to be almost exactly above the Earth's North Pole, the solar system's north pole is actually 23 degrees away, near the star Zeta Draconis. Neither one is close to our galaxy's north pole, which lies in the direction of the star 31 Comae Berenices, halfway between Spica and the Big Dipper's bowl. The sky reveals the result of these different angles: the zodiac (the plane of the solar system) is angled 60 degrees to the Milky Way Galaxy.

Moving farther out, the Andromeda Galaxy's north pole is tipped 12 degrees compared with our galaxy. There seems to be no grand scheme of cosmic angles. Even on the largest scale, galaxies are grouped in clusters that are apparently at random tilts to each other.

THE SUN AND THE MOON

Multiple exposures of the total eclipse of the sun on February 26, 1979, taken about 10 minutes apart, show the sun partially eclipsed both before and after totality (when the sun is completely covered by the moon). The Earth's rotation caused the shift in position from one exposure to the next. Photograph by William Sterne.

An obscure documentary film made in the mid-1980s has a remarkable sequence that begins with interviews with graduates from Harvard University immediately after the cap-and-gown ceremony. During each interview, the graduates are asked whether they would mind answering a few general-knowledge questions. One of them is: Why is it warmer in summer than in winter? Almost all of them, science and arts graduates alike, say that it's because Earth is closer to the sun during summer. That's wrong, of course. The answer is that in summer in the northern hemisphere, the tilt of the Earth's axis angles this hemisphere so that (a) the sun is higher in the sky, and its rays shine down more directly, and (b) daylight lasts longer. But few of the graduates knew that.

It seems astonishing that a famous university, noted for its scholars, had just graduated people who didn't have a grasp of something as fundamental as the seasons. However, I have no doubt that

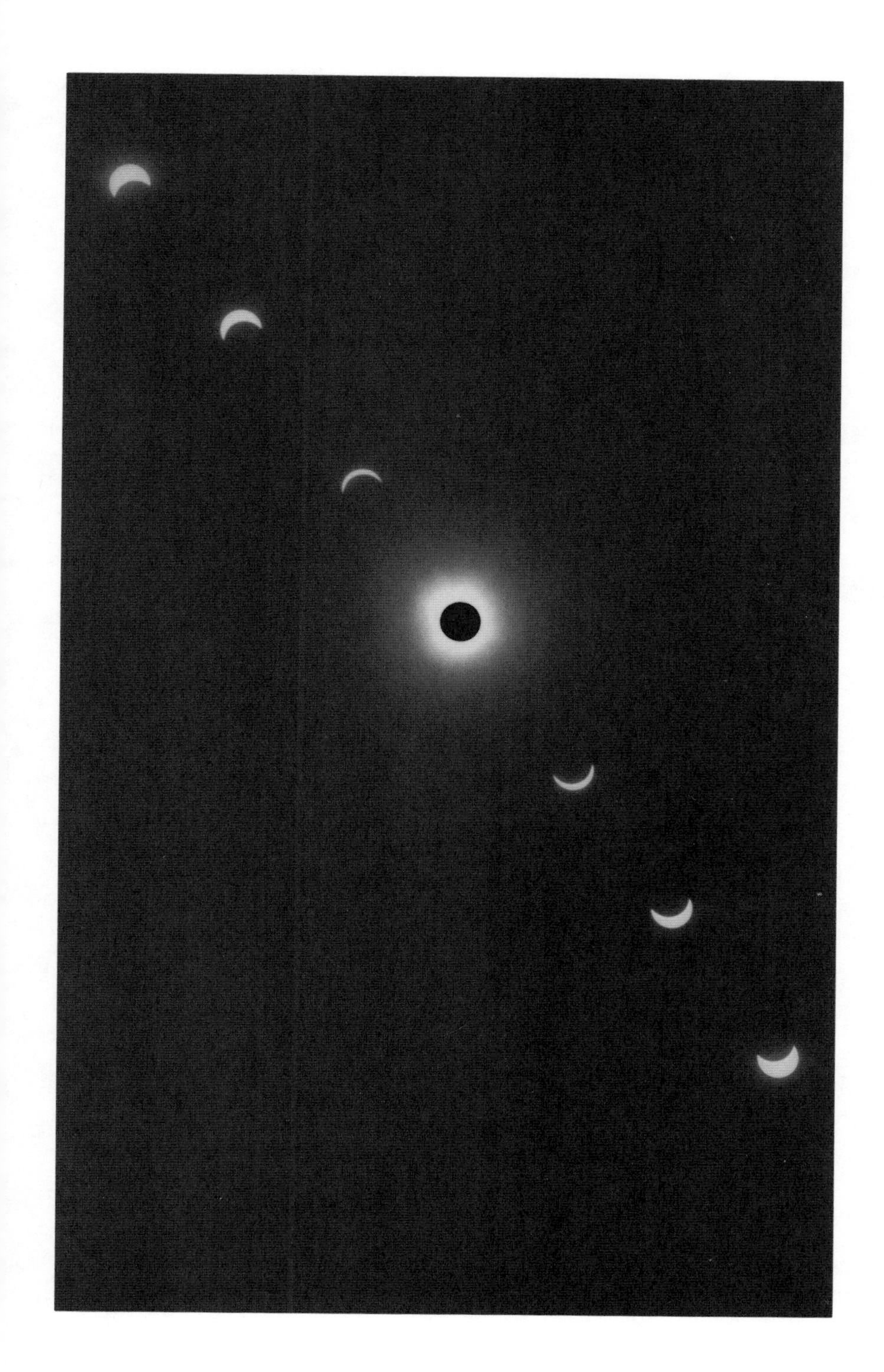

graduates of any university would have given the same answers, so Harvard is not the culprit here.

A colleague of mine who teaches an astronomy course at a less well known university told me that he tested his own students at the end of the course, and well over half of them could not explain why the moon has phases. Yet most of them had passed the term exam, which included questions about Kepler's laws of planetary motion and the redshift of galaxies. The course has since been refocused.

What has gone awry here in epidemic fashion is, I think, the generally abstract context in which so many subjects are presented—throughout schooling, not just university. Connections to the real world of nature wither in the face of too much chalkboard and textbook and not enough observation of everyday phenomena as simple as the altering length of shadows and changes in the duration of daylight from season to season.

One of the most crucial life skills a person can develop is the ability not just to see but to *observe*. I know there are exceptional teachers who concentrate on imparting this skill. I wish there were more.

30 *CAN A TELESCOPE SHOW WHERE THE ASTRONAUTS LANDED ON THE MOON?*

A backyard astronomer's telescope is able to reveal amazing detail. The craters and ancient lava plains of our nearest celestial neighbour—the moon—stand out so distinctly that it seems like a view through a spacecraft window. The vista is stark and impressive, but most people looking at the moon's alien landscape for the first time are lost for a sense of scale. Having no idea how big the craters and other features are, observers naturally wonder whether they can see the lunar landers and other equipment left on the moon.

The moon is a substantial world in its own right, fully one-quarter the diameter of Earth. Its biggest craters are typically 50 to 100 kilometres across. The smallest craters visible in large amateur-astronomy telescopes are about two kilometres wide. But although the general locations of the Apollo landing sites can be identified using a moon map, the lunar

Six pairs of American astronauts landed on the moon between 1969 and 1972 during the climax of the Apollo programme. Each team left a lunar lander and other paraphernalia that will remain essentially unchanged until explorers return to our nearest neighbour world in the centuries ahead. NASA photograph.

lander and other equipment left behind by the astronauts are much too small to detect, even with the largest telescopes on Earth.

31 IS THE MOON MOVING CLOSER TO OR FARTHER FROM EARTH?

Astronomers calculate that the moon is receding from us at the rate of three centimetres per year, pinpointing the cause as tidal friction. The moon's gravitational attraction pulls on Earth, which produces tidal bulges on opposite sides of our planet. As Earth rotates, the planet's oceans rise and fall, sloshing along the continental coasts. Such movement creates friction, which slows the Earth's rotation in the same way that the brakes of a car slow the rotation of its wheels.

The slowing of the Earth's rotation is very slight, amounting to a lengthening of the day by one second every 62,500 years—an apparently infinitesimal amount, but it adds up. In four million years, a day will be one minute longer. One billion years from now, a day will be about 28 hours long. During the Age of Dinosaurs, a day was only 23 hours long.

This very gradual slowing of the Earth's spin

means a continuous loss of energy that is equivalent to two billion horsepower. However, the energy, in the form of angular momentum, has to go somewhere, and a lot of it is transferred to the moon. Added energy of motion moves the moon to a higher orbit, which pushes it farther from Earth, just as a burn from the space shuttle's engines would do.

This action could continue for billions of years until Earth slows enough for it to end up with one side always facing the moon. Then the Earth's day and the moon's orbital period will be the same, about 47 days. However, a study conducted at NASA's Jet Propulsion Laboratory suggests that about two billion years from now, the moon's orbital period and the Earth's rotation will become tangled in a series of what are known as spin-orbit resonances, which could dramatically alter the tilt of the Earth's axis. After that, it is difficult to predict the Earth-moon system's evolution.

The last-quarter moon as seen in a 100mm refractor telescope, typical backyard astronomers' equipment. Photograph by Terence Dickinson.

32 *WHY DOES THE MOON LOOK SO BIG WHEN IT IS RISING?*

There is no doubt about it: the moon appears larger when it is close to the horizon than when it is near overhead. Yet you can prove to yourself that this is strictly an illusion. Hold an aspirin tablet at arm's length in front of the moon. Regardless of the moon's elevation in the sky, the aspirin will just cover it.

But I can recommend no experiment which will alter the fact that the moon (and the sun too) *appears* so much bigger when it is near the horizon. Refraction, or lensing, by the Earth's atmosphere is not responsible. The refractive properties of the atmosphere compress the moon's disc, rather than bloat it. This can be seen most clearly at sunset as the "oval" sun dips below the horizon.

The truth is, this effect is entirely conjured up by the human brain. One of the most powerful illusions in nature, it is also one of the least understood. Numerous scientific studies have grappled with the exact causes of the horizon illusion, or moon illusion, as the effect is sometimes called, but no single explanation is widely accepted. However, the leading theory goes something like this:

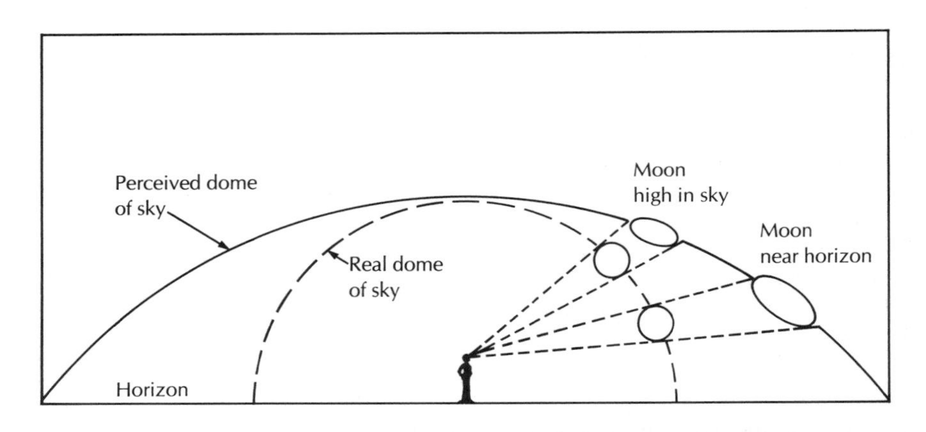

The moon illusion—the apparent enlargement of the moon when it is near the horizon—is one of the most powerful optical deceptions in nature, and it is still not fully understood. The moon does not change in diameter; it is the way we perceive the moon's image that causes the effect. Much of the illusion is thought to centre on the human perception of the sky as an overturned bowl, more distant at the horizon than overhead. But, as explained in the text, it is more complicated than that. Photograph by Terence Dickinson.

First, the brain is apparently "prewired" to assume that objects high in the sky are closer than objects near the horizon. This is probably because clouds are, in fact, closer when overhead than when on the horizon. Hence, the theory claims, the brain subconsciously tells itself that the moon is more distant on the horizon. In reality, however, the image of the moon that registers on the retina is the same size as when it is overhead, as you proved with the aspirin tablet. But in order for the moon to appear the same size at a greater distance, the brain reasons, it must be physically larger on the horizon. And the brain goes one step further by enlarging the image to fit the "logic." And that enlargement is what is sent to the conscious mind.

If you find that explanation unconvincing, I don't blame you. But the other theories are even less plausible. The phenomenon remains a deeply mysterious eye/brain illusion. Yet there is no denying that it *is* an illusion. Knowing that your brain is sending you the wrong signals doesn't help either. I still see the full-blown effect. And so will you.

I am sure that psychologists in the 21st century will still be trying to understand the horizon/moon illusion. Meanwhile, here is a final test to prove that it's all in the mind. When no one is watching and the full moon is near the horizon, bend over and look at the moon from between your legs. The effect vanishes, presumably because you have flipped the scene reaching the brain by placing the horizon above the moon.

This may sound confusing, and it is. Since the roots of the illusion reach back through millions of years of evolution, we are dealing with a collective human memory that once must have served some useful purpose. Today, it is a puzzling artifact that affects us all.

33 *WHY IS THE MOON'S DARK SIDE SOMETIMES VISIBLE AT CRESCENT MOON?*

Occasionally, when the moon is a thin crescent, the rest of it is dimly visible as a ghostly glow. This phenomenon, which is especially beautiful through binoculars, has been described as "the old moon in the new moon's arms." It is known to astronomers as Earthshine and is simply sunlight reflected off Earth onto the nighttime side of the moon.

To visualize Earthshine more easily, imagine that you are an astronaut on the night side of the moon. There is no sun in the sky, but Earth looms as a brilliant blue-and-white globe in the blackness—a beautiful celestial lamp brightening the landscape.

Faint illumination of the crescent moon's night side is a beautiful sight in deep twilight. The phenomenon, known as Earthshine, is strikingly enhanced when viewed through binoculars. Photograph by Terence Dickinson.

Earthshine illuminates the lunar night 50 times more brightly than the moon lights our nights, which is bright enough to make the night side visible to observers on Earth. Leonardo da Vinci was the first to deduce that it is actually light reflected from Earth which makes visible the part of the moon not in sunlight.

Earthshine varies in intensity by up to 15 percent depending on the presence of differing amounts of cloud cover on the side of our planet facing the moon, but it is most intense when the moon is a thin crescent. At that time, our satellite is being illuminated by a nearly "full" Earth.

In the northern hemisphere, the most favourable time to watch for Earthshine is two to four days after the new moon in February, March, April, May and early June, when, as a result of seasonal changes in celestial geometry, the crescent is higher in the sky than at other times of the year.

34

WHAT IS A BLUE MOON?

Every 2.7 years, on average, there are two full moons in a month. The second one has come to be known as the blue moon, as in the saying "once in a blue moon." Future blue moons occur on July 30, 1996, and March 31, 1999.

In researching this question, however, I could find no historical basis for the definition of a blue moon as the second full moon in a month. Although the first mention of a blue moon goes back to 16th-century playwrights William Roy and Jerome Barlowe, neither they nor anyone else until relatively modern times had anything to say about two full moons in a month. I have concluded that the two-full-moons-in-one-month definition of the blue moon is a recent invention.

For centuries, the coincidence of calendar and celestial clockwork that produces two full moons in a month was ignored by astronomers and everyone else. It was no big deal. Then, in the past few decades, newscasters and weather forecasters began to claim that the second full moon in a month is the blue moon. It is a modern myth.

Yet the expression "once in a blue moon" must have some basis in nature. The explanation I prefer is one given by the late Helen Sawyer Hogg, a Canadian astronomer. In the 1960s, Hogg pointed out that blue moons—and blue suns—really do occur in nature, albeit rarely. They are caused by the dust created by major volcanic eruptions and by the smoke from forest fires, the type of dust and smoke that is often made up of particles slightly larger than a wavelength of light. These particles act as a filter, because they scatter red light but do not impede blue light, which has a smaller wavelength. As a result, only blue light reaches the eye.

"There seems little doubt that the well-known blue-moon saying is derived from such circumstances," Hogg wrote in 1975. "From a study I made of the literature of the past several centuries, it appears that volcanic eruptions and other effects will render a blue moon visible probably once in seven or eight decades"—roughly once in a lifetime.

The last widely observed blue moon was seen in late September 1950, following a series of huge forest fires in northern Alberta. The region of blue-moon visibility stretched from Canada to Florida and east to northern Europe. Astronomy historian Patrick Moore observed the phenomenon from England. On September 26, 1950, at East Grinstead, Sussex, Moore made this entry in his observing log: "The moon shone . . . with a lovely shimmering blueness—like an electric glimmer—utterly different from anything I have seen before." Moore collected dozens of newspaper accounts of the event from across England and from other parts of Europe, noting that most of the reports did not make the connection with the forest fires.

The conditions that produce a real blue moon seem to be so rare, few people have ever seen one.

35 WHAT CAUSES HALOES AROUND THE SUN AND THE MOON?

Ring around the moon,
Rain will come soon.

Such snippets of weather lore are usually only partly right, and this one is no exception. Although lunar haloes often precede rain or snow by a day or two, I'd rather put my money on the official forecast.

One thing those old rhymes *do* tell us is that several generations ago, people were very aware of changes in the sky. When is the last time you saw a halo ringing the sun or the moon? Have you *ever* seen one? The phenomenon is actually quite common; probably 20 lunar haloes and 40 solar ones are visible each year from any one location.

Haloes around the sun may go unnoticed, however, because the sky is often extra bright when a halo is present. The brightness is caused by white cirrostratus clouds, which give rise to the halo effect. Cirrostratus are thin, milky sheets of cloud that veil the sky but allow the sun to shine through. These clouds are made of tiny prismatic ice crystals that tend to refract, or bend, sunlight at one specific angle, 22 degrees, which creates a halo 44 degrees

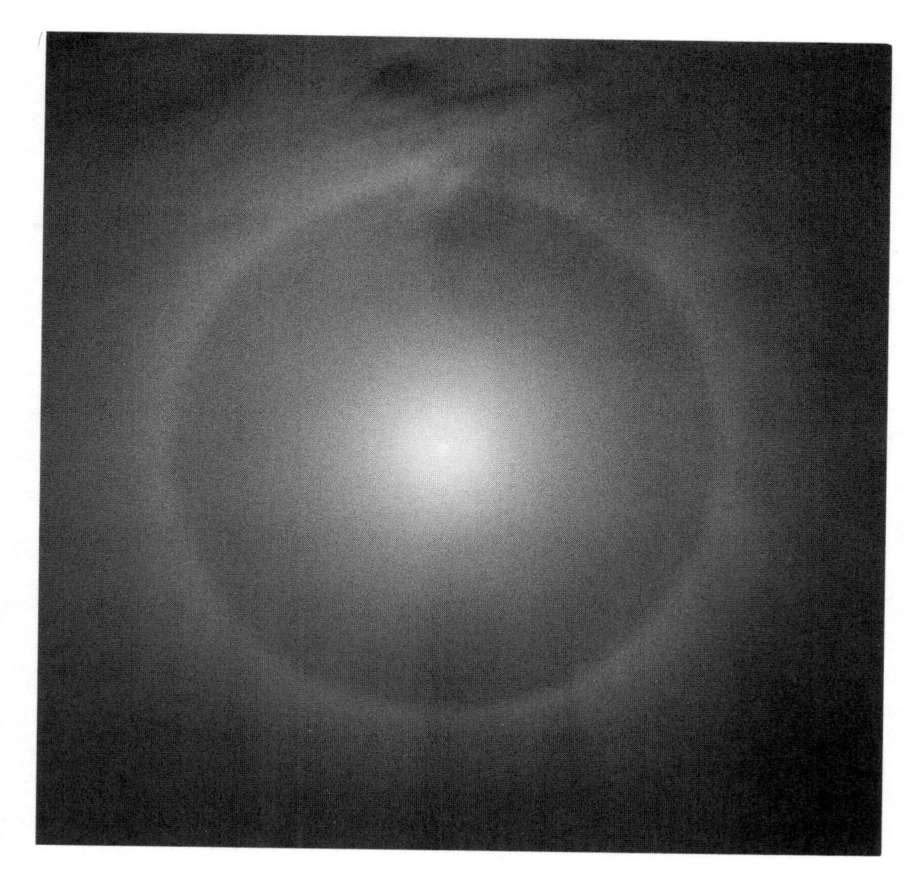

A lunar halo is caused by the refraction of moonlight through ice crystals in thin cirrus clouds. While quite common, both lunar and solar haloes are often missed by people who seldom look skyward. Photograph by Alan Dyer.

in diameter—about half the angular distance from the horizon to overhead.

Solar haloes are most often seen in cooler weather, when conditions are ideal for the formation of the best type of cirrostratus clouds. Since haloes are big, one way to get a good look at them is to stand just within the shadow of a tree or a building that will block direct sunlight and thereby reveal half or more of the halo.

Cirrostratus cloud formations also create haloes around the moon. Lunar haloes are often more conspicuous than solar haloes because they are seen against a dark night sky. The best time to look for both types is in calm weather, when the temperature is just above freezing.

A related sky phenomenon is the sundog, a brilliant glow seen well to the right or left of the sun when it is fairly low in the sky. The six-sided pencil-shaped ice crystals that make up cirrus or cirrostratus clouds can float in the air straight up and down, refracting sunlight and thereby creating a sundog.

If the sundog is especially bright, its colours will be obvious—red on the side closest to the sun, blue on the far edge and yellow in the middle. Sundogs are usually brightest in winter because the tiny ice crystals are more common then, but they can be seen all year if you look for them. That's the trick—deliberately watching for nature's displays.

36

WHEN IS THE NEXT ECLIPSE?

There are two very different kinds of eclipses—solar and lunar—and two types of each. Here's a summary:

A ***total solar eclipse***, by far the most spectacular eclipse, occurs when the sun is completely covered by the moon. Think of it this way: To view a total solar eclipse, you must stand in the moon's shadow. Although the moon is one-quarter of the Earth's diameter, its shadow, where it strikes our planet, is only about 100 kilometres wide. As the moon moves in its orbit, the shadow sweeps across the daytime side of Earth at roughly 1,500 kilometres per hour. Astronomers call the track of the moon's shadow the path of totality. Amateur astronomers, naturalists and eclipse buffs travel the far reaches of the globe to see a total solar eclipse, even though the phenomenon never lasts longer than seven minutes. They say one exposure to the splendour of totality is addictive, and I must agree. The quick plunge into darkness as the moon's shadow sweeps past and the sudden appearance of bright stars, plus the sun's delicate pearly corona and brilliant red prominences, all add up to an unforgettable experience.

A ***partial solar eclipse*** is seen over a vastly larger region of Earth than the path of totality. But the difference between a total and a partial solar eclipse is literally the difference between night and day. Even if 99 percent of the sun is covered, it still shines forth as it would on a heavily overcast day, and therefore, proper filtration for direct viewing is required. A total solar eclipse rates a 10 on a 1-to-10

Partial eclipses of the sun, in which the moon covers only a portion of the solar disc, are visible several times a decade from any particular place on Earth. The rarest partial eclipse —an annular eclipse— occurs when the moon, near its maximum distance from Earth, does not fully obscure the sun. The result is a solar ring, above. Photographs by Terence Dickinson (left) and Ron Schmidli (above).

scale; a partial solar eclipse gets a 3. There are two ways to view a partial solar eclipse in complete safety. The first is to punch a small hole in a piece of cardboard with a pencil point and use it as a pinhole projector that will throw an image of the sun's disc on a white card held a foot or so behind. The second is to buy a No. 14 welders' filter plate, available at many welding-supply outlets for a few dollars. *Do not use substitutes.* Techniques using smoked glass, dark bottles or layers of exposed film do not block the infrared radiation from the sun that can damage your eyesight. (By the way, it is dangerous to look at the sun at *any* time; no additional radiation is emitted during an eclipse.)

A ***total lunar eclipse*** occurs when Earth casts its shadow on the moon. In this case, the shadow is big enough to engulf the moon completely, with room to spare. The portion of the eclipse when the moon is entirely inside the shadow (totality) can last for

The radiant corona—the sun's atmosphere—is brilliantly displayed when the moon covers the sun during a total solar eclipse. No protective eyewear is required to see the corona during totality. Photograph by Robert May.

1 to 1½ hours and is visible anywhere on the side of our planet facing the moon. The moon is not blacked out by the shadow, because red light from the sun filters through the Earth's atmosphere and is refracted into the shadow, causing the moon to dim to a dull copper, brown or grey hue. (Once or twice a century, the shadow is dark enough to make the moon disappear at mid-eclipse.) A total lunar eclipse is readily visible with no optical aid. I give it a 7 on our celestial-splendour scale.

A ***partial lunar eclipse*** occurs when the moon swings through the edge of the Earth's shadow. It is only slightly more common than a total lunar eclipse but is less impressive, because the subtle shading of the eclipsed region is often overwhelmed by the brilliance of the portion of the moon still in full sunlight. I rate it 3 or 4 on our scale.

FUTURE TOTAL SOLAR ECLIPSES

1994, November 3: South America, South Atlantic
1995, October 24: India, Southeast Asia, Indonesia
1997, March 9: Russia
1998, February 26: Pacific, South America, Caribbean
1999, August 11: Europe, Middle East, India
2001, June 21: South Atlantic, Africa

FUTURE PARTIAL SOLAR ECLIPSES VISIBLE FROM NORTH AMERICA

1994, May 10: *major event*; sun more than 80% covered for much of the United States and eastern Canada; best solar eclipse of any type visible from North America until 2017
1995, April 29: Central America and Caribbean only
1998, February 26: south of a line from San Diego to Milwaukee
2000, December 25: northeastern North America
2001, December 14: western North America
2002, June 10: western North America

FUTURE TOTAL LUNAR ECLIPSES VISIBLE FROM NORTH AMERICA

1993, November 28: entire continent
1996, April 3: eastern half of North America
1996, September 26: entire continent
2000, January 20: entire continent
2003, May 15: entire continent
2003, November 8: entire continent except West Coast
2004, October 27: entire continent

FUTURE PARTIAL LUNAR ECLIPSES VISIBLE FROM NORTH AMERICA

1994, May 24: entire continent except Alaska, Yukon and northern British Columbia
1997, March 24: entire continent
1999, July 28: early-morning event for West Coast only

As the first table reveals, we are in a total-solar-eclipse drought in North America. The next total eclipse of the sun over Canada or the United States is a long way off, on August 21, 2017. Then the path of totality will stretch from Oregon to central Nebraska and on through Tennessee and South Carolina. Less than seven years later, on April 8, 2024, another cross-continent total solar eclipse occurs, this one running through central Texas, Arkansas, Indiana, Ohio, southern Ontario and southern Quebec.

STARGAZING

Few sights in nature can match the majesty and inspiration of a star-filled night sky. But because most of us live in or near cities, the stars are seldom seen in their proper glory. Only during camping trips or other ventures far from urban life do the stars shine forth. Paradoxically, astronomy has never been more popular. Amateur-astronomy clubs report regular increases in membership. More telescopes were sold in the past decade than during any previous 10-year period. Astronomy books are selling better than ever.

Recreational astronomy is a fast-growing hobby that can carry the mind of the beholder on a soaring journey through the cosmos. Photograph by Terence Dickinson.

Perhaps it is the well-known phenomenon that we often don't appreciate things until we have almost lost them. Stargazing (or recreational astronomy, as amateur astronomers prefer to call it) has become somewhat exotic, even to the point that some individuals buy fancy, gadget-laden telescopes more as showpieces for their den than as tools to explore the cosmos. Conversely, nothing more than a pair of binoculars is needed to enjoy astronomy.

If you have ever thought of expanding your interest in astronomy to include actually locating and observing some of the celestial objects that you read about on these pages and elsewhere, you should find some useful tips in this chapter.

37

WHAT IS THE EASIEST WAY TO LEARN THE CONSTELLATIONS?

After 35 years of testing what does and doesn't work for stargazers trying to learn to identify the constellations, I can offer a few guidelines. First, some things to avoid:

Don't bother with any device that has glow-in-the-dark star maps. Although it may sound like a good idea, such charts are just a jumble of dots under the stars. Also, stay away from star-finding devices that you look into or sight through. I have tried them all and have never found one that helps a true beginner.

Another basic rule is to resist the temptation to buy a telescope before you know what the hobby is about. A telescope is much easier to use *after* you learn the constellations, rather than before. With their jiggly mounts and inadequate instruction manuals (usually poor translations from Japanese), such telescopes are not suitable introductions to astronomy (see Question 44).

What you should do is buy one or two guidebooks that have clear, simplified star charts. (I immodestly suggest my own book *NightWatch*.) Unless you are fortunate enough to have an amateur-astronomer friend who will point out the constellations, this is the only proven way to find them. To start off, there are a few steps that require no equipment and just two basic charts. Using these, you will be well on your way to exploring the universe. Both charts show only bright stars, so they can be used in urban, suburban or country settings.

Chart A is centred on the Big Dipper. This chart can be used any night of the year in Canada and the United States (except autumn evenings in the southern states). Many people are able to recognize the Big Dipper as soon as they step outside, but those who have trouble are usually looking for something smaller than the Big Dipper, which is a bit wider

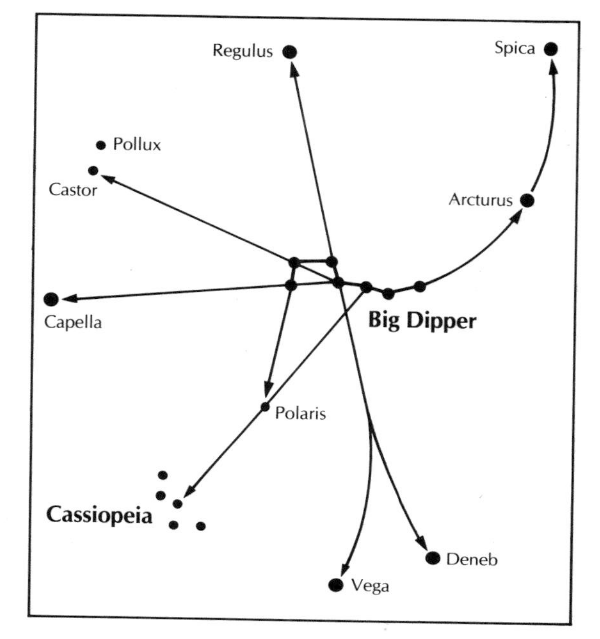

Chart A: The Big Dipper is the key to the night sky for observers in the northern hemisphere. Its seven bright stars point the way to at least nine other prominent stars and constellations. This chart is useful any night of the year throughout Canada and the United States. Because of the Earth's varying position in its orbit, the Big Dipper can be angled differently than shown, but it always retains its utility as a guidepost.

than the distance from your thumb to your little finger when your hand is spread open and held at arm's length. One source of difficulty in identifying the Big Dipper is the fact that it can be oriented at any angle, depending on the season or the time of night. But one thing is constant—it is always in the northern part of the sky.

Once the Big Dipper is sighted, use the locater arrows to identify a few of the brightest surrounding stars. Not every star on Chart A is visible at the same time, but over the course of a year, they are all prominent for two or three seasons.

Chart B is centred around Orion, the brightest of the constellations. Orion is a winter star group as seen from the northern hemisphere, visible in the southern evening sky from late November to early April. Look for Orion's distinctive three-star belt. The whole constellation, top to bottom, is slightly smaller than the Big Dipper. The configuration even looks somewhat like a stick figure of the mythological hunter after which the constellation was named. That is unusual; most constellations do not look like their namesakes. Many of them were named for figures prominent in mythology rather than for a

Chart B: Orion, the mighty hunter of Greek mythology, is the night sky's brightest constellation and the chief star finder after the Big Dipper. From the northern hemisphere, Orion is seen in the southern sky from late November to mid-April. About the same dimensions as the Big Dipper, Orion sports a three-star belt that makes the constellation unmistakable. All the stars marked on this chart and on the chart on the previous page are bright enough to be seen from a large city.

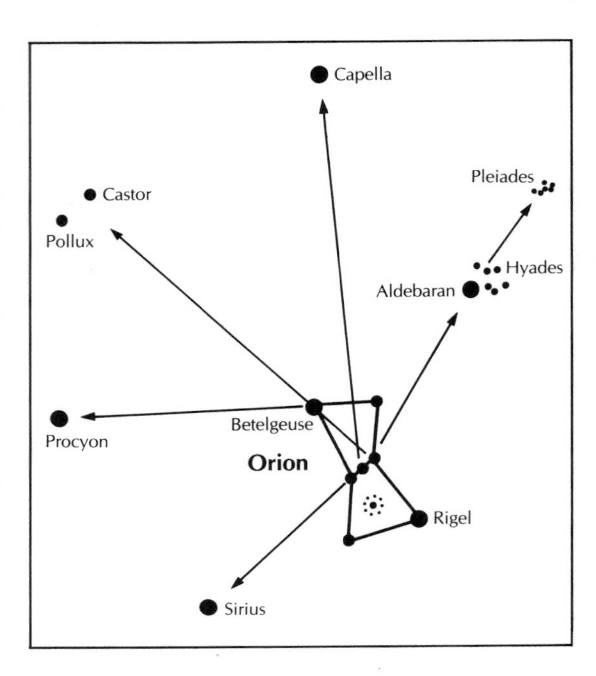

close fit with any connect-the-dots pattern. For this reason, I discourage learning too many constellation shapes initially. Sighting prominent stars using locater arrows is the easiest way to begin.

38 HOW MANY STARS ARE VISIBLE ON A CLEAR, DARK NIGHT?

The night sky deep in the country may appear to be plastered with millions of stars, but under the best conditions, most people will see only about 4,000. It seems like a serious miscount, doesn't it? But you can prove it to yourself. The next time you are under a sky with "countless" stars, try this experiment:

Touch your thumb to your forefinger to form a circle, like an okay sign. Hold the circle about a foot in front of one eye, and close the other. Count the stars inside the circle. Then try another spot, and count again. Three or four counts should be enough. The total for each count could be as high as 40 stars but will more likely be in the 5 to10 range. Take an average, and then multiply by 200 to get an estimate of the sky's total cargo of visible stars.

Above: The brilliant bulge in the Milky Way marks the centre of our own galaxy.
Left: A close-up view of the densest region of the Milky Way Galaxy, revealing thousands of individual stars. Photographs by Terence Dickinson.

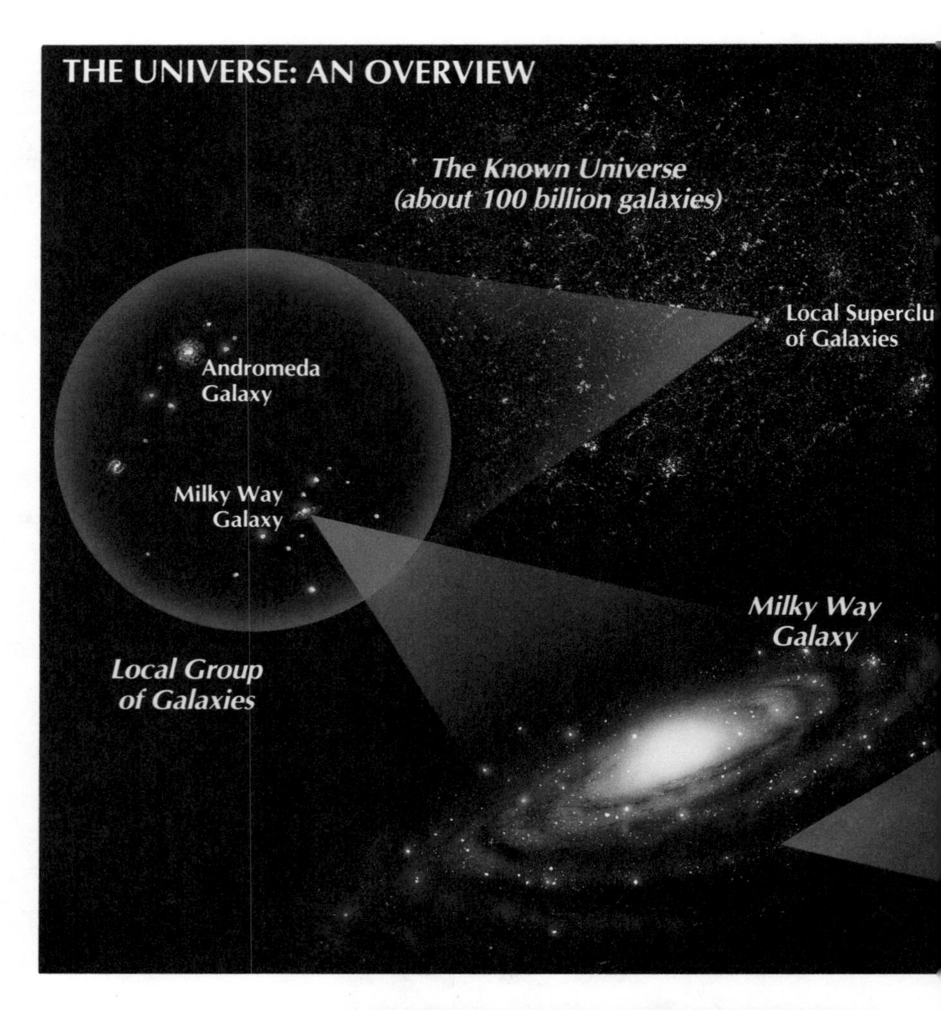

Above: The Earth is one of nine planets orbiting the sun. The sun is one of several billion stars in the Milky Way Galaxy. The Milky Way Galaxy is, in turn, one of billions of galaxies in the universe. Illustration by John Bianchi. Right: The Andromeda Galaxy. Photograph by Terence Dickinson.

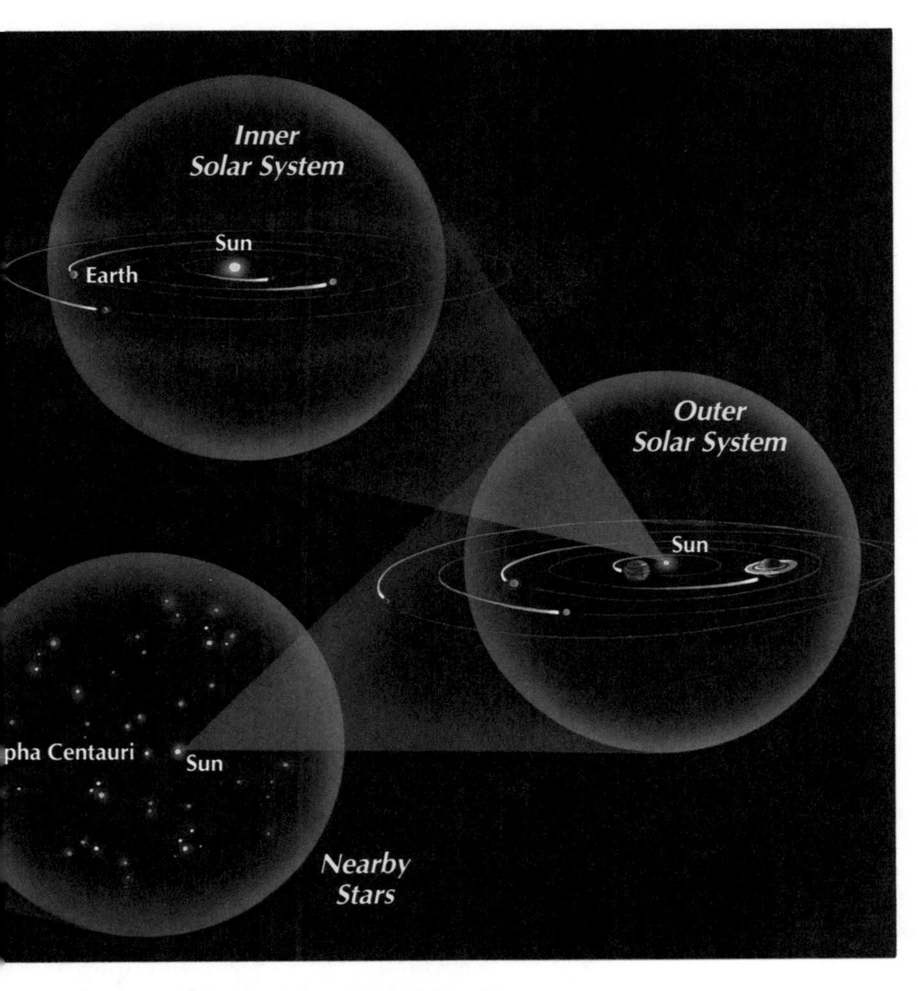

Left: Seen edge on from a distance of 30 million light-years, the Milky Way Galaxy would probably look much like the galaxy NGC 891 that we view from the same distance. The dark obscuring nebulas in NGC 891 are similar to the dark rifts visible in the Milky Way. Photograph by Jack Newton.

Above: A spectacular display of aurora borealis washed over much of North America on November 8, 1991. Right: The December 9, 1992, total eclipse of the moon was visible throughout the Americas. Facing page: A seven-hour time exposure of star trails over the University of Toronto Southern Observatory on Las Campanas, Chile. Photographs by Terence Dickinson.

Above: Only during a total eclipse of the sun can the delicate pearly solar corona and ruby tongues of gaseous fire—the prominences—be seen in their true glory. Photograph by Jerry Lodriguss.
Right: Less than 12 years after Sputnik 1, the first astronauts landed on the moon. NASA photograph.

Above: Jupiter is a colossal ball of gas topped by swirling clouds spun into belts by the planet's rapid rotation. The famous Great Red Spot is at left. Photograph by Terence Dickinson and Alan Dyer, using the University of Toronto 24-inch telescope on Las Campanas, Chile.
Left: Mars photograph by the Hubble Space Telescope.

Above: The elegance and exquisite beauty of Saturn's rings are captured in this 1990 Hubble Space Telescope image, the best ever obtained from Earth. Right: The Eta Carinae Nebula, a 200-light-year-wide cloud of celestial gas and dust illuminated by stars within it. Photograph by Terence Dickinson.

The six brightest members of the Pleiades star cluster are visible to the unaided eye under dark skies. Dozens of fainter ones can be picked up with binoculars. The cluster's stars are about 450 light-years from Earth. Photograph by Leo Henzl.

The factor 200 is roughly the number of thumb-and-finger circles it would take to survey the visible night sky. The number of *potentially* visible stars is more than double that total because Earth is in the way and blocks half the cosmos at any particular time. Another consideration is horizon haze, which reduces the tally still further. If these two effects are excluded, the total rises to about 9,000 stars.

Contrary to some reports of astronaut views of the stars from space, an astronaut well away from Earth does not see significantly more stars than an earth-bound observer. According to Jay Apt, an American astronaut who knows the constellations and took the time on a space shuttle flight in 1991 to do the comparison, the view from Earth at a perfectly dark site far from city lights is very similar to the view of the same section of sky from space.

39 HOW FAR OUTSIDE THE CITY DO I NEED TO GO TO SEE THE STARS PROPERLY?

Here are the things I look for when selecting a site for stargazing:

Direct Light. Any lights that shine unimpeded onto the site become a severe distraction as your eyes attempt to adapt to the darkness. Streetlights and security lights are the worst offenders, because they are on from dusk to dawn, but porch lights and ballpark lighting can also be an astronomer's nemesis. Finding a spot without such interference—or being able to use trees or buildings to block it out—is the first priority.

Indirect Light. Often called light pollution, indirect light is the sky glow seen over cities every night. It is the air being illuminated by outdoor lights, and the effect is growing year by year (once installed, streetlights are virtually never removed). Light pollution could be significantly reduced with better-designed lighting fixtures that direct the light where it is needed. But regardless of its cause, sky glow makes the night in and around cities a permanent twilight that blocks the splendour of the starry night from view. If you can see the Milky Way, you have escaped the worst of it. At a really good site, the Milky Way is a distinct silver ribbon visible most of the way to the horizon.

Personal Security. Standing outside at night, alone and in an unfamiliar setting, may be the best way to see the stars, but there is always the possibility of the unexpected. Reduce the chances of having a stargazing session interrupted in an unfriendly way by observing from a known stargazing site, such as an astronomy club's outpost or a state or provincial park. Parks for campers and naturalists are often the safest and darkest accessible locations, and many have scheduled stargazing sessions with telescopes provided.

40 IS IT POSSIBLE TO PHOTOGRAPH THE STARS WITH A REGULAR CAMERA?

Modern cameras have become so automated that many models cannot be set for the lengthy time exposures necessary for starlit skies. Unless you can turn off the auto features so that the camera can be used manually for time exposures longer than 10 seconds, it will not work for star pictures.

If your camera can be set for time exposures, you will also need a camera tripod and high-speed film (400 speed is fine, but my favourite is Kodak Ektar

1000). Use a lens of 50mm or shorter focal length.

On a dark, moonless night, with the camera mounted on a tripod and pointed at the Big Dipper or some other easily recognized grouping, set the lens at f/2.8 or f/4, and cap the lens. Lock open the shutter. Remove the cover for 15 to 30 seconds, then replace it and release the shutter. With this simple technique, your pictures will reveal all the stars that are visible to the naked eye. Ektar 1000 will show even more. Try it.

What about video cameras? Most video cameras can be used to photograph the moon through a telescope. Simply aim the camera lens into the telescope eyepiece, and record what you see in the camera's viewfinder. Some recent video camera models have remarkable low-light capacity and will reveal features in an almost darkened room. These

The Big Dipper is at right in this moonlit winter scene taken from eastern Canada. The two outermost stars of the Big Dipper's bowl point to Polaris, the North Star, one dipper-length away. Pictures such as this can be taken with any camera loaded with high-speed film and capable of exposures of more than 15 seconds. Photograph by Terence Dickinson.

cameras can record stars when aimed at the night sky, but not as effectively as the conventional high-speed-film method mentioned above. Specialized still-video cameras, called CCD-imaging cameras, have amazing capabilities and are becoming widely used by amateur astronomers, although they cannot be applied to home-movie use.

41 WHEN IS THE BEST TIME TO SEE METEORS ("SHOOTING STARS")?

Meteors zip across the sky every night of the year at the rate of 5 to 10 per hour. Typically the size of a peanut, a meteor incinerates as it plunges into the Earth's atmosphere at about 50 times the speed of a rifle bullet. The heated air particles along the path give the appearance of a streak that can last a few seconds, in the case of a bright meteor. Meteors seldom penetrate below 10 kilometres in altitude.

Most meteors are comet debris, pebblelike particles that were once mixed in the ice which forms the bulk of a comet. When a comet reaches the inner solar system, its surface is vaporized by the sun's heat. Solid pieces that have been trapped in the comet's ice since the formation of the solar system 4.6 billion years ago are released and begin to travel around the sun in the same path as the comet. The comet's orbit eventually becomes a river of particles that can survive for centuries.

Over time, the meteoric particles disperse from their parent comet's orbit and blend into a background of debris that produces the sporadic meteors visible on clear, moonless nights. However, for several nights each year, Earth encounters streams of meteors that have not yet dispersed fully. During these passes, the number of meteors streaking across the sky increases to 30 or more per hour, and we are treated to a meteor shower.

The two most productive meteor showers are the Perseids (usually August 11-12) and the Geminids (mid-December). Dust and debris shed by Comet Swift-Tuttle cause the Perseids, while the Geminids are the remnants of an object called Phaethon, thought to be an extinct comet that no longer releases much gas or dust. Because the orbits of these

Up to 10 meteors per hour slash the sky every night of the year from a particular location. But this number increases dramatically during the two major annual meteor showers, the Perseids (August 11-12) and the Geminids (December 13-14). This meteor was seen during the 1990 Geminid shower, when the author counted more than 200 meteors during a four-hour observing session. Photograph by Terence Dickinson.

objects cross the Earth's orbit, we intercept the meteor streams at roughly the same time every year.

Swift-Tuttle is an active comet with a 134-year orbit that carries it out past Pluto. A recent visitor to the inner solar system, Comet Swift-Tuttle swung through the portion of its orbit that intersects the Earth's orbit in December 1992. The added material from this pass is expected to produce excellent displays of Perseid meteors over the next few years (for related information, see Question 46).

42 WHEN IS THE BEST TIME TO LOOK FOR THE NORTHERN LIGHTS?

Two or three times a decade, the entire night sky is alive with dancing, luminous curtains of red, green, purple and yellow light, swaying and pulsating and apparently converging near the zenith (overhead).

Experiencing such a display of northern lights, or aurora borealis, is like standing inside nature's kaleidoscope. On these occasions, the rare all-sky aurora is at maximum strength—trillions of watts of nature's power surge through the high atmosphere, creating an unforgettable spectacle.

Although bright enough to be seen from within a large city, all-sky auroras have been witnessed by few people. The fact is, our indoor television-oriented culture in combination with modern outdoor lighting mean that many people have never seen an aurora at all. Admittedly, dazzling all-sky auroras are relatively infrequent. But lesser displays are visible several times a year from midnorthern latitudes. In 1991, for example, I saw modest auroras on 42 different nights from my home just north of Lake Ontario (latitude +44 degrees).

Auroras peak and subside in harmony with the 11-year sunspot cycle. The last sunspot maximum was 1991, and spectacular all-sky auroras were visible in March 1989, March 1990 and November 1991. Now that we are past the maximum, auroras are less frequent and less intense, although an impressive show can still occur. Specific predictions are impossible. There is no favoured time of night, although the strongest displays seem to occur in March, April, September and October. The only way to see an aurora is simply to go outside every clear night and look for an unusual brightening in the northern sky.

Ranging from a pale greenish white glow near the horizon to intense red, green and purple spears and curtains that fill the sky, auroras magically float among the stars. The phenomenon originates when eruptions on the sun's surface, called solar flares, liberate vast amounts of charged particles into space. The charged particles—actually just parts of normal atoms—reach Earth, follow our planet's natural magnetic field and are funnelled into a continent-sized ring around the magnetic north pole in Canada's Arctic (a similar ring occurs over Antarctica).

The interaction between air molecules 100 to 800 kilometres above the surface and the incoming solar-charged particles releases energy in the form of light, making the Earth's upper atmosphere act like a glowing television screen. This is happening all the

The most intense displays of aurora borealis, or northern lights, produce brilliant gossamer curtains visible over huge areas of North America and Europe. This one was seen from Edmonton, Alberta. Photograph by Russ Sampson.

time, but usually, it is dim or the auroral ring is too far north to be visible from the more populated parts of Canada and from the United States. When solar activity increases and more particles reach Earth, the aurora brightens, the ring expands to the south and the display may become visible to millions.

43 ARE WINTER NIGHTS CLEARER AND DARKER THAN SUMMER NIGHTS?

When I first learned to identify the constellations, it was the middle of winter. I remember noticing on warm evenings a few months later how much duller the sky looked. The stars do indeed appear brighter in winter, but it has nothing to do with the crisp air. Astronomers have tested air clarity and found no difference among the best nights at a specific site regardless of the season.

But there is some truth to the notion that the stars are different in winter. They twinkle more vigorously and somehow seem brighter, as if invigorated by the cold air. The effect is most noticeable on frigid nights, when dry snow crunches underfoot.

However, the real reason the stars look brighter on a winter night is that evening skies in winter contain *more* bright stars than in other seasons. With more bright stars, the sky naturally appears more dazzling, thus the illusion that the cold air is responsible—but it isn't.

Here are the specifics: The most prominent winter constellations—Orion, Gemini, Auriga, Taurus, Canis Major and Canis Minor—contain 17 of the 33 brightest stars visible from midnorthern latitudes. Yet these six constellations are collected in just one-tenth of the whole sky. The key to winter-sky brightness is that the only time all of the six constellations decorate the evening sky is from mid-December to late March, which coincides almost exactly with the coldest weather in the northern hemisphere. In other seasons, the Earth's orbital motion around the sun turns us to face less starry parts of the galaxy.

Why are there more bright stars in one part of the sky? It is not entirely random clumping. About half of those 17 bright stars were born during the last few million years from a vast gas and dust cloud, a neb-

Night skies in winter often seem to be clearer and more star-filled than at other times of the year. But are they really? Photograph by Terence Dickinson.

ula, about 1,000 light-years away and centred in Orion. They are also among the most luminous stars in the Milky Way Galaxy, ranging up to 150,000 times the brightness of the sun.

Superluminous stars like those that form the outlines of the winter constellations are exceedingly rare. For every star 100,000 times as bright as the sun, there are 1,000 stars like the sun and 50,000 stars fainter than the sun. That statistic implies there is more up there than meets the eye. Throngs of fainter stars are utterly invisible to unaided eyes. Good binoculars will begin to reveal the hordes. A telescope shows more.

44 WHAT TYPE OF TELESCOPE SHOULD I BUY?

Frankly, most of the telescopes available in camera and department stores and Christmas catalogues are a waste of money. Almost every telescope below $350 falls into this category. I wish it were otherwise, but these telescopes are made and packaged for impulse purchase, often as gifts. Complete with fancy dials, multiple eyepieces and shiny knobs, they look impressive, but it's a façade—all show and

Recreational astronomy can be a rewarding hobby when teamed with a quality telescope such as this 80mm refractor by Celestron (about $500).

no substance. Once set up, they tend to be flimsy, jiggly and frustrating to aim and focus.

Especially beware of any "beginner's" astronomical telescope that prominently displays the instrument's high magnification, since high power makes a telescope difficult to aim and keep on target. In any case, most celestial objects are best seen at moderate magnification (20 to 100 power) in small telescopes. And don't be swayed by well-known brand names either. Regardless of whose name is on them, these so-called beginners' instruments are all made in the same two or three factories in the Orient.

What, then, do I suggest?

In a word, binoculars.

Binoculars are the best starter instrument you can own. They have many advantages, the most important of which are modest price and an inviting wide, bright field of view that can be seen with both eyes. Conversely, inexpensive beginners' telescopes almost always have narrow, dim fields that make observing more difficult than it should be. I recommend 7 x 50, 10 x 50, 7 x 42 and 8 x 40 sizes, although other binocular sizes will do almost as well. (The first number is magnification; the second, the diameter of the two main lenses in millimetres.)

An important binocular accessory is a tripod

So-called beginners' telescopes with spindly mounts and inadequate optics are readily available in most camera and department stores. See answer to Question 44 for guidelines on what to look for and what to avoid in a first telescope.

adapter that allows you to attach standard binoculars to a camera tripod. If the binoculars have a threaded hole at the front of the focusing bar (sometimes covered by a small screw-on cap), the adapter is a simple $15 L-shaped bracket. Alternative clamp brackets are available for many binoculars without an adapter hole. The advantage of the tripod mount is that it eliminates the shakes of hand-held viewing. The difference in clarity is amazing.

Although I rant that most beginners' telescopes are trashy, there are a few models I like, but you will probably find them only in telescope specialty stores (check the Yellow Pages under "Telescopes") or in mail-order advertisements in *Astronomy* and *Sky & Telescope* magazines. Recommended models (all $300 to $700) are: Edmund's 4-inch Astroscan; any 80mm or 90mm refractor telescope by Meade, Vixen, Orion or Celestron; Coulter's 8-inch or 10-inch Dobsonian; Celestron's C4.5 or SP-C6 Newtonian reflector; Meade's Starfinder 6-inch Newtonian reflector.

For more details, see the chapter on equipment for backyard astronomers in my book *NightWatch* or write for the excellent booklet *Buying Your First Telescope*, available for one dollar from *Astronomy* magazine, Box 1612, Waukesha, WI 53187.

ALIENS AND MORE . . .

The Orion Nebula is a looming presence in the night sky of a hypothetical Earthlike planet in a distant solar system. Perhaps the most compelling mystery in science is the question of the Earth's uniqueness. Are we alone in a universe of a billion trillion suns? Illustration by John Mitchener.

Our final category is for questions that do not readily fit into what has been covered so far: a mixed bag of observatories, rocks from the sky, a famous star in the Bible and musings about extraterrestrial life.

Questions about life elsewhere in the universe are, and always have been, of central importance to humanity's probes into the cosmos—whether by mind, eye, telescope or spacecraft. In the fourth century B.C., the Greek philosopher Metrodorus of Chios wrote: "To consider Earth the only populated world in infinite space is as absurd as to assert that in an entire field sown with seed, only one grain will grow."

Metrodorus was the first known thinker to speculate about extraterrestrial life, although he was undoubtedly preceded by thousands of others who gazed at the night sky and wondered what the stars were trying to tell them. However, not until the second half of the 20th century have we been able to do much more than speculate. Now, radio tele-

scopes seek signals from our cosmic cousins and space probes sniff the air and poke the soil of other worlds to try to detect signs of life.

45 HOW BIG IS THE WORLD'S LARGEST TELESCOPE?

The giant reflector telescope atop Palomar Mountain, near Escondido in southern California, is the best-known telescope on Earth. For decades, it was also the largest. Completed soon after World War II, the 200-inch telescope dwarfed all other astronomical instruments in existence. Today, half a century later, the Palomar telescope is still one of the best astronomical research tools on Earth.

Rivals to the 200-inch were built during the 1970s, but only one exceeded it in size, and none significantly outperformed it. Astronomers realized that traditional telescope designs had reached their limit. They were simply too massive to work effectively in sizes much beyond 200 inches. Tests showed that owing to its enormous weight, a 300-inch telescope would be unacceptably bent by gravity (the 200-inch telescope weighs 500 tons; a 300-inch of the same design would be more than 1,000 tons). Some way had to be found to reduce the weight of the huge primary mirror made of special quartz glass. Likewise, the gigantic mounts had to be simplified, perhaps by turning the complex tracking duties over to computers.

During the 1980s, new designs were perfected that would make very large telescopes much lighter. The colossal 10-metre (394-inch) Keck telescope, completed in 1992, is the first of the new-generation telescopes. Like all the world's large telescopes, the Keck instrument is a reflector, but its main mirror is not a single piece of glass. Instead, it is a mosaic of 36 smaller, specially shaped pieces designed to fit snugly together into a single giant. Each of the segments is computer-controlled by positioning pistons to keep it aligned with all its neighbours, so the parts act as a whole. The result is a huge segmented mirror with about one-tenth the weight of a similar one in the old Palomar design.

Other large telescopes scheduled for completion

THE WORLD'S LARGEST TELESCOPES

Telescope/Nation	*Location*	*Size* (m)	*Size* (in.)	*Completed*
Keck Telescope/U.S.	Mauna Kea, Hawaii	10	394	1992
*Keck II Telescope/U.S.	Mauna Kea, Hawaii	10	394	(1997)
*Columbus Telescope/U.S.-Italy	Mt. Graham, Arizona	8.4	330	(1996)
*Subaru Telescope/Japan	Mauna Kea, Hawaii	8.3	325	(1997)
*European Very Large Telescope†	Cerro Paranal, Chile	8.2	320	(1996)
*Gemini/U.S.-U.K.-Canada	Mauna Kea, Hawaii	8.1	315	(1998)
*Magellan Telescope/U.S.	Las Campanas, Chile	6.5	255	(1997)
*MMT Observatory/U.S.	Mt. Hopkins, Arizona	6.5	255	(1994)
Russian Astrophysical Observatory	Caucasus Mountains	6.0	236	1976
Palomar Observatory/U.S.	Palomar Mt., Calif.	5.1	200	1949
William Herschel Telescope/U.K.	Canary Islands	4.2	165	1986
Cerro Tololo/U.S.	Cerro Tololo, Chile	4.0	156	1975
Anglo-Australian/U.K.-Aust.	Siding Spring, Aust.	3.9	153	1975
Kitt Peak National Observatory/U.S.	Kitt Peak, Arizona	3.8	150	1974
U.K. Infrared Telescope	Mauna Kea, Hawaii	3.8	150	1979
European Southern Observatory	La Silla, Chile	3.6	142	1976
Canada-France-Hawaii Telescope	Mauna Kea, Hawaii	3.6	141	1979

*planned or under construction; projected completion date shown in parentheses
†the VLT will eventually be an array of four 8.2-metre telescopes side by side

The Keck telescope, a 10-metre reflector and the world's largest, is the first of a series of new-generation telescopes under construction during the 1990s. Photograph by Roger Ressmeyer/Starlight for the California Association for Research in Astronomy.

For decades after its completion in 1949, the 200-inch Hale telescope on Palomar Mountain in California was by far the biggest and most powerful on Earth. Larger new-generation instruments (see table, previous page) use less bulky computer-controlled mounts and mirror supports. Caltech photograph.

in the 1990s use thin single-piece mirrors, rather than the segmented solution. Although these mirrors warp as the telescope moves, modern computers can completely correct for this in a way that the designers of the Palomar telescope could not have imagined. The table lists the world's 10 largest telescopes (as of 1993), along with 7 others in the planning stages or under construction.

46 HOW OFTEN DO METEORS REACH THE GROUND?

At 9 o'clock on the morning of September 29, 1938, on a quiet street in Benld, Illinois, a meteorite the size of a grapefruit smashed through a garage roof, then through the roof of the car inside, bounced off the car floorboards and became lodged in the seat upholstery. A witness said it sounded like an airplane in a power dive, but no flaming object was seen, and the meteorite was not hot enough to singe the ripped upholstery.

Nearly half a century later, on the evening of November 8, 1982, members of a family in Wethersfield, Connecticut, were watching television when they were shocked by a crash that sounded as if "a truck [was] coming through the front door." Rushing to the front room, they saw a hole in the ceiling. The intruder, a football-sized meteorite, had rammed through the roof and ceiling, hit the floor and rolled into the dining room. A jogger eight kilometres away had seen a fireball overhead that "easily cast shadows." He noted that the dazzling object had appeared nearly motionless in the sky. "It was a head-on look," he realized later. Then the object had disappeared. He did not see it descend toward Wethersfield, nor did he hear it crash through the roof, even though he was facing the town. But 30 to 50 seconds later, he heard what sounded like "rifle shots" from the direction of Wethersfield. The sounds were small sonic booms from the descending meteorite.

Incidents like the Wethersfield and Benld impacts are exceptionally rare. Only a few dozen cases of damage to a building have been documented. No one in recorded history has ever been killed by a

meteorite, although at least two people have been injured when they were struck by baseball-sized rocks from space. While the chances of seeing a meteorite reach the ground are extremely remote, many people imagine they have witnessed just that, because a brilliant meteor slashing toward the horizon can give the impression it has landed a few city blocks away. That is almost never the case. For one thing, the average bright meteor is only about the size of a peanut and completely burns up between 50 and 100 kilometres above Earth.

Very bright meteors, called fireballs, range from the size of a walnut to that of a baseball or larger—big enough to plunge to lower levels and appear like chunks of molten or incandescent metal hurtling earthward. A few reach the surface. The best estimates suggest that on average, 50 meteorites larger than a plum hit our planet each day, although only a handful are ever recovered. They are made of rock or metal and originate in the asteroid belt beyond Mars. Parts of rockets and satellites add to the total, as they fall from orbit almost daily. Larger chunks sometimes reach the ground. Whenever possible, Space Agency controllers direct an incoming satellite to plunge into the southern region of the Pacific or Indian Ocean, where ships seldom travel.

On Friday, October 9, 1992, at 7:50 p.m., 18-year-old student Michelle Knapp was at home in Peekskill, New York, when she heard a loud noise "that shook the windows." She rushed outside and discovered a washtub-sized hole in the trunk of her 1980 Chevy Malibu. Underneath the car, sitting in a small crater, was a rock about the size and shape of a football. As word of the event spread, Knapp was contacted by collectors eager to buy the 12.4-kilogram meteorite. She happily accepted $69,000 for the cosmic intruder. Photograph by John Bortle.

If a meteoric chunk survives to fewer than 10 kilometres above the Earth's surface, it stops glowing as it is significantly decelerated by atmospheric drag and falls to Earth like a stone dropped from an airplane—fast, but at nowhere near the 50,000-kilometre-per-hour velocity it has at the top of the atmosphere. That is why the jogger did not see a flaming object plunge into the town. It also means there is no way of knowing whether an object you might see will actually survive to the Earth's surface. Chances are, it will not, unless the initial fireball appears much brighter than the brightest stars.

Note: The terms used in this field of science can be frustrating at first. A meteor *is the bright flash seen in the sky, not the object that causes it. A* meteoroid *is any small body in space—the stuff that creates a meteor when it plunges through the atmosphere. A* meteorite *is any chunk that survives to the Earth's surface.* Meteoritics *is the study of all three of the above.* Meteorology *is the study of none of the above; it is the science of weather forecasting.*

47 WHAT WAS THE STAR OF BETHLEHEM MENTIONED IN THE BIBLE?

No celestial phenomenon has generated so much debate for so long as has the Christmas Star. Astronomers from Johannes Kepler to Carl Sagan have pondered the star's origin, but the meagre details in the Bible permit many interpretations. All the evidence pivots on a single New Testament account, Matthew 2:2, which relates that astrologers (Magi or Wise Men) from the East (modern-day Iraq) journeyed to Jerusalem asking, "Where is he that is born King of the Jews? for we have seen his star in the east, and have come to worship him."

Despite the tradition of Christmas illustrations, the New Testament states neither how many Magi there were nor what the star looked like. But we do get some additional information (Matthew 2:9): ". . . and, lo, the star, which they saw in the east, went before them, till it came and stood over where the young child was."

Most biblical scholars suggest that the twice-

The Star of Bethlehem as depicted in a 15th-century German woodcut.

repeated phrase "in the east" refers to where the Magi were when they saw the star, rather than the direction in which they were looking. The thrust of modern research has been to identify a celestial phenomenon that the Magi would have interpreted as associated with the Jews and with the birth of a leader. The Magi were on the lookout because of a prediction in the Old Testament (Numbers 24:17): ". . . there shall come a Star out of Jacob, and a Sceptre shall rise out of Israel."

The most enduring theory is the Jupiter-Saturn triple conjunction of 7 B.C., when the two planets passed within two moon diameters of each other in May, September and December. A similar Jupiter-Saturn triple conjunction occurred in 1981, but such events are rare enough to be limited to one or two per lifetime. Although a triple conjunction is of only casual interest today, ancient astrologers such as the Magi would have been intrigued, because two of the most important cosmic objects were apparently in conference in the constellation Pisces, which was at that time associated with the Hebrews.

The thinking today is that the second of the three conjunctions would have convinced the Magi to embark on their journey, the third occurring when they were near Bethlehem. In this scenario, Jesus was born in late 7 B.C. or early 6 B.C. This date fits with statements in *Antiquities of the Jews* by Flavius Josephus (37-95 A.D.), considered to be a reliable historian of the day. Josephus notes that King Herod died just before Passover and just after a lunar eclipse. If the partial eclipse of March 13, 4 B.C., is selected, everything falls into place. (The Bible implies that Jesus was 1 or 2 years old when Herod died.)

However, there was a much more impressive total eclipse of the moon on January 9, 1 B.C. If this is the one associated with Herod's death, the birth of Jesus (and the star) occurred in 2 or 3 B.C. The only significant celestial event in this time frame that would have interested the Magi occurred on June 17, 2 B.C., when Jupiter and Venus produced a spectacular conjunction, appearing so close that they were seen as a single object. Furthermore, it happened just as darkness fell in the Middle East. At dusk, the two planets appeared very close together, almost touching. An hour later, they seemed to merge into one brilliant object.

The Venus-Jupiter conjunction took place in the constellation Leo, between the front and back feet of the traditional lion as pictured among the stars. Astronomer Roger Sinnott notes that a Jewish religious amulet dating from the first few centuries A.D. depicts a lion with a star between its feet. The lion's ancient association with kings, he argues, makes Leo an obvious constellation for the Star of Bethlehem.

The one element that does not fit this theory is the biblical passage from Matthew about how the star "came and stood over where the young child was," seeming to imply that the star moved. Modern retranslations suggest the meaning might be that the star was seen again over Bethlehem. In either case, the triple-conjunction interpretation is more plausible, since the Magi could have embarked after the second conjunction and observed the third, 11 weeks later, over Bethlehem.

Regardless of the details, all experts agree that the word "star" in the Bible could also refer to a planetary conjunction, a comet or some other celestial phenomenon. Halley's Comet, an obvious potential contender, appeared in 12 B.C. and is therefore out of the running.

What about another comet or a nova or a supernova, the explosive brightening of a star? Apparently, nothing exceptional appeared. We know this because Chinese astrologers kept meticulous records throughout the time span in question, and those records still exist. The conjunction theories are the only ones that fit the facts as we know them today, although there are still enough imponderables to allow debate over the Star of Bethlehem for a few more centuries.

48 *ARE UFOS REAL?*

If I had to pick one question that I have been asked most often over the past 30 years, this is probably it. My answer is yes—a qualified yes.

UFO means unidentified flying object, and certainly, I believe there have been many legitimate sightings of aerial phenomena that have remained baffling and, hence, unidentified to the observer. This is an important distinction. Unidentified to the observer is not the same as not identifiable. Nor do reports of such sightings, even when they number in the thousands, necessarily mean that we are being visited by extraterrestrials, which is what I think most people are asking when they pose the question.

Here is an example that will illustrate my position.

In 1973, I was teaching an astronomy class at the Strasenburgh Planetarium in Rochester, New York. While we were on the top-floor observatory deck identifying constellations, we saw a formation of lights pass silently almost overhead, then veer off and swoop toward the horizon. It was not impossible for an aircraft or a group of aircraft to behave in this manner, but I had never before seen such a formation nor the type of lights we saw that night.

What made this sighting baffling was the call I made the next day to the Air Force base responsible for the airspace over Rochester. The community relations officer at the base denied that any Air Force vehicles were anywhere near the area at the time of our sighting. Through persistent sleuthing, I finally concluded a few weeks later that we had seen the powerful downward lights of a squad of helicopters, which were apparently a military secret at the time.

The giant robot Gort and the alien Klaatu emerge from a flying saucer in the classic 1951 science fiction film The Day the Earth Stood Still. *This was the first Hollywood film to treat the subject of extraterrestrial intelligent life seriously. It also firmly established the flying saucer as the mode of transport for visitors from the stars. Photograph courtesy Twentieth Century Fox.*

This interpretation was later confirmed by an astronomy professor at a nearby college, who had actually observed the craft by accident through the college observatory telescope.

My point is that without the time and effort taken to research this case properly, I would have been left with a UFO story to tell for the rest of my life. The incident prompted me to look at the subject more closely. In the years following, I investigated many UFO sightings and was, for a time, a consultant for a major UFO study. There were a few interesting cases (such as the advertising plane with marquee-style moving lights; it made a terrific UFO), but I soon realized that something as familiar to me as the planet Venus or the star Sirius could be so alien to some people, they would call the police or airport control tower for an explanation.

One evening, for instance, our local television news anchorman seemed genuinely excited by a UFO sighting that he and a cameraman investigated. "And we have from the scene, perhaps for the first time anywhere, live footage of a UFO," he announced, barely able to maintain his practised earnest and unflappable demeanour.

The poor fellows had videotaped the planet Venus, which became an enigmatic dancing dot through the lens of the camera perched on the cameraman's shoulder. There it was, mysteriously bobbing in and out of focus as the auto-focus video camera struggled with its impossible task. After the news broadcast, I called the anchorman. It was clear from the time and direction involved, I told him, that Venus was the object in question. He accepted my explanation, albeit reluctantly.

But what about the hard-core cases that can't be explained?

I am more than ready to be convinced that just one of these UFOs is a device piloted by or controlled by extraterrestrials. I would like to be convinced. I have no problem with the idea that intelligence from other planets could visit Earth. I would love to be alive when humanity learns that extraterrestrial intelligence exists.

While current UFO evidence may be enough to persuade some, however, it is too flimsy to persuade me. Anecdotes supported by ambiguous evidence are not good enough no matter how much we might want them to be true. UFO investigations have revealed the amazing variations in the way people perceive their environment, but not much else. The UFO data might even hold some profound secrets of life on other worlds, but I don't believe it does. After nearly half a century of UFO sightings, we have learned remarkably little about the phenomenon.

Finally, there is a simple fact: No one spends more time looking at the night sky than amateur astronomers. (Professional research astronomers attached spectrographs and electronic cameras to their telescopes long ago and seldom need to look through them or at the starry night.) Yet for all their time under the stars, amateur astronomers see remarkably few UFOs. They see satellites crashing to Earth, exploding stars and millions of aircraft lights but not

An alleged flying saucer photographed over the island of Trinidad in 1958 is typical of hundreds of pictures of UFOs published since the term "flying saucer" was coined in 1947. None can be taken as proof of visits by extraterrestrials, and many have been shown to be lens reflections or simply fakes.

the close-encounter UFO stuff. After thousands of hours of night-sky observing, I have still never seen a UFO. Why not? Shouldn't knowledgeable sky observers be the first, rather than the last, to witness the phenomenon? It just doesn't add up.

More on this in the following question.

49 WHY DO ASTRONOMERS SEARCH DISTANT STARS FOR EXTRATERRESTRIAL SIGNALS BUT INSIST THAT EXTRATERRESTRIALS CAN'T BE NEAR EARTH?

This question illuminates what I believe is a serious contradiction in the statements made by many who recommend the use of radio telescopes to search for extraterrestrial intelligence, an endeavour known by the acronym SETI (pronounced SET-ee).

SETI advocates point out that in a galaxy like ours, one star in a million might have a planet very similar to Earth orbiting it. In this scenario, our galaxy—the Milky Way—which has more than 300 billion stars, would have several hundred thousand Earth-like planets. The most efficient way for civilizations on those planets to communicate with other civilizations, say SETI advocates, would be by signals that travel at the speed of light. Radio waves do this and have the additional advantage of punching through nebulas that block light waves. Furthermore, radio transmitters are easy to build and inexpensive to operate and can transmit and receive over vast distances through space. An alien civilization may be beaming signals at Earth even as you read this. If we could detect such a signal, it would prove that intelligent extraterrestrials do exist. And if we don't listen, we will never know.

One could quibble about the details in this scenario, but most would argue that searching for extraterrestrial signals is at least worth a try. NASA, the U.S. Space Agency, agrees, and in October 1992, powerful signal-analyzing computers were hooked up to radio telescopes in California and Puerto Rico to begin a 10-year $100 million search of the galaxy.

Where the reasoning breaks down, in my view—and herein lies the contradiction—is the contention

This astronomical radio telescope in Algonquin Park, in Ontario, was briefly used to search for signals from extraterrestrials in the 1970s. During the 1990s, NASA is funding a much more extensive search using similar radio telescopes in California and Puerto Rico. Photograph by Terence Dickinson.

by SETI advocates that it is exceedingly unlikely that extraterrestrials have visited Earth or that they could even if they wanted to. On the one hand, they suggest that the galaxy is teeming with life; on the other, they imply that all those civilizations would be cocooning—staying at home, content to communicate by radio signals or just to sit and listen in order to discover who is out there.

"Space travel is slow and enormously expensive," say the SETI searchers. "Radio communication is cheap and works at light speed." They buttress these statements with calculations showing the inefficiency and impracticality of spaceships. "Spaceflight between the stars would be done, but not often," one SETI pioneer allowed in answer to my direct question on this point.

To me, such arguments are specious. They remind me of remarks made by Harvard University astronomer William H. Pickering in 1910. He predicted that it would never be possible to build aircraft engines powerful enough to propel passenger airplanes across the Atlantic. And, of course, he completely dismissed any notion of travel to the moon. Yet he was one of the few astronomers of his day to suggest that there was evidence of some form of life on the moon.

For me, the central question is: Why would inquiring creatures be satisfied to communicate for millennia by megaphone? Clearly, the *only* way to explore effectively is to go there. Let me explain.

Fossil evidence shows that life has been on Earth for at least 3.8 billion years, more than three-quarters of the time the planet has existed. But only in the past few decades has it been possible for this life to reach beyond its home planet by spacecraft and radio signals. Extraterrestrials have had 3.8 billion years to discover life on Earth by coming here and having a look but only 60 years to detect our radio chatter from afar. Which has more potential for discovery, exploring or staying home?

Direct exploration, even by robot devices, would reveal life forms no matter what their stage of development. Radio searches may be fine for us at this point in our evolution, but to assume that such searches are the preferred mode of operation for civilizations millions of years in advance of us sounds suspiciously like a repetition of the uninspired pronouncements of William Pickering more than 80 years ago.

Why is this illogical scenario repeated so often by prominent SETI advocates? I believe the main reason is, they want to distance themselves as far as possible from the UFO advocates who claim that extraterrestrials are already here. By focusing the extraterrestrial question on the radio search, the subject embraces a high-tech setting—computers, radio telescopes, star charts, and so on—rather than the maze of anecdotes, misinterpretations and sensationalism that surrounds much of UFO research, at least the part encountered by the public.

While I can't blame anyone for doing this and have no objection to the radio searches, I think it is more reasonable to assume that long ago, advanced extraterrestrials discovered that life exists on Earth. They have known about us for aeons and are capable of travelling here whenever they choose. Perhaps the fact that we remain unaware of their existence is more a reflection of their priorities than ours.

50 *WHAT ABOUT THE STORIES OF PEOPLE BEING ABDUCTED BY ALIENS?*

It sounds like a supermarket-tabloid creation: Hundreds of people claim to have been taken against their will into flying saucers, where they were examined by aliens and released, virtually unharmed. The "victims," according to researchers in the field, are not publicity seekers. Yet publicity about the abduction phenomenon has never been greater than in recent years. Clearly, there is widespread interest in such bizarre experiences. In my estimation, if there is even the slightest possibility that any of the claims are true, they are worthy of all this attention—and they should receive more, including close scientific scrutiny for information about extraterrestrial life.

That is, of course, *if* the abduction stories can be taken at face value.

According to psychiatrists who have studied them, abductees are generally otherwise normal people with no history of psychosis or mental disease. However, the memories of the abduction are often repressed, and in these cases, hypnosis is usually required to draw out details of the incident. The abductees describe the alien creatures in very similar terms—about four feet tall with greyish skin, large heads and huge black eyes—and report undergoing the same types of physical examinations.

"The amazing part is the similarity of detail in these accounts," says Temple University historian David Jacobs, who has taken a special interest in the cases. "There is no known explanation for it. Indeed, it defies rational explanation."

A few years ago, I was involved in a television special about the abductee phenomenon, and I witnessed the hypnosis of an abductee. Rather than becoming more convinced, however, I became more skeptical. To my mind, the abductee was too will-

ing, too susceptible to the friendly and sometimes predictable questions of the hypnotist.

I have also read much of the literature on this subject and have interviewed several abductees and the recognized experts in the field. Frankly, it all seems a bit too flimsy. For one thing, if the events described by these individuals and in the books on the subject are real, they flatly defy the laws of physics. These alien creatures can enter buildings without tripping sophisticated burglar alarms, walk through walls, levitate themselves and their human subjects, communicate thoughts without speaking, partially paralyze humans at will and induce amnesia of the event.

What bothers me most, though, is that all the "evidence" is anecdotal. Anything presented as physical evidence is always ambiguous. For instance, many abductees claim to have had tiny crystal-like devices painfully implanted in their sinus cavities behind the bridge of the nose. Investigators have examined these people and even resorted to CAT scans, all with negative or inconclusive results.

Even novelist Whitley Strieber, an abductee who wrote a best-seller based on his own abduction experiences, says he "by no means accepts the extraterrestrial explanation." Neither do I. Robert Baker, a professor of psychology at the University of Kentucky, goes even further, noting that "Strieber shows the classic symptoms of a fantasy-prone individual:

Most drawings of alien beings that are based on descriptions by "abductees" show creatures four to five feet tall, with greyish skin, large heads and huge, dark eyes. Why so many people who don't know each other claim to have seen the same thing is a mystery. But can this be regarded as evidence of visits by extraterrestrials?

He is easily hypnotized, he is amnesiac, he has vivid memories of his early life, he has a religious background, and he is a writer of occult and highly imaginative novels." Baker dismisses Strieber's alien encounters as "textbook hypnopompic hallucinations that can happen to perfectly normal people but are much more common with fantasy-prone people."

Why are the abductees' stories so similar? Psychologist Milton Rosenberg of the University of Chicago suggests that UFO abduction tales represent a "pervasive modern myth." As the myth spreads, through supermarket tabloids, "nonfiction" best-sellers and television "documentary" shows, says Rosenberg, more and more people recast their own fantasies or hallucinations in terms of the myth.

Rosenberg's suggestion may be completely wrong, but there are many skeptics with similar opinions. As I said earlier, I am ready to be convinced by an unambiguous piece of solid physical evidence of extraterrestrial creatures or their technology. Not stories. Evidence.

Until the evidence gap changes, the explanation of the abduction phenomenon in my view remains entirely earthbound, the apparent product of human, rather than alien, minds. While I have sympathy for the abductees, who seem to suffer very real stress and torment, I can't believe that this is how we are going to find out about our cosmic cousins.

APPENDIX
THE AMAZING UNIVERSE

I conclude our explorations with this random compilation that further celebrates the extraordinary diversity of the cosmos.

The star PSR1937 + 21 is no wider than the city limits of a town the size of Lincoln, Nebraska, or Kingston, Ontario, but it weighs 500,000 times as much as Earth and is spinning on its axis 642 times per second. This king of the class of twirling stars called ***pulsars*** is spinning so fast that its equator is moving at one-tenth the speed of light.

Pulsars are incredibly dense. A thimbleful of pulsar material would weigh as much as all the water in Lake Erie. This density means that a pulsar's gravitational attraction at its surface is far beyond anything we experience within the solar system. If a spaceship could land on a pulsar, it would be instantly crushed into a pool of subatomic particles about the thickness of the nucleus of an atom.

The most distant known objects in the universe are ***quasars***—most likely the dazzling energetic cores of colliding galaxies—at an estimated 12 billion to 14 billion light-years from Earth. The light we see from them began its journey 12 billion to 14 billion years ago, 7 billion to 9 billion years before Earth even existed.

About 99 percent of the ***stars*** seen on a typical night are larger, more massive and brighter than the sun. Each one of them is moving at about 400,000 kilometres per hour, cruising in an orbit centred on the Milky Way Galaxy's nucleus.

Europa, one of Jupiter's four largest moons, may be the best place in the solar system beyond Earth to look for life. When the Voyager 2 spacecraft swept by Europa in July 1979, it revealed a world encased in ice. Scientists later realized that they were looking at the frozen surface of a global ocean—of water. Estimates of the ice thickness vary from one or two kilometres to tens of kilometres, but the ocean below, kept liquid by Europa's internal heat, might support some form of life.

On the night of February 23-24, 1987, at Las Campanas Observatory, Chile, Canadian astronomer Ian Shelton was developing a photograph he had taken earlier that night of the Large Magellanic Cloud, a satellite galaxy to our Milky Way. Shelton immediately noticed that the picture showed a brilliant star where none should be. He rushed outside, looked up and saw that the star was real, not a flaw in the film. It was the discovery of the brightest ***supernova*** seen from Earth since 1604.

The ***Great Red Spot***, a storm cell in the atmosphere of Jupiter, is three times larger than Earth. If an object the size of Earth collided with Jupiter, the event would be merely a flesh wound for the planetary colossus. Its cloud patterns and atmospheric wind belts might be disrupted for a few centuries, but no permanent effects would remain.

If a particle the size of a baseball in ***Saturn's rings*** were brought back to Earth and placed on a table, it would turn into a puddle of mud within a few minutes. Ring particles are mainly ice with dirt mixed in.

Saturn is primarily light gases such as hydrogen and helium. The entire planet has the same average density as softwood. If there were an ocean big enough, Saturn could float in it.

The ***galaxy M87***, visible in a small telescope, contains an estimated 10 trillion stars. Each one of those stars is at least 25,000 times more massive than Earth.

A ***black hole*** the mass of Earth would be smaller than a golf ball. A black hole the mass of the sun would be no wider than the city limits of a town of 10,000 inhabitants. A black hole the mass of the entire galaxy would be 10 times wider than the solar system.

Not until the late 1920s—a mere three generations ago—did astronomers finally gather enough evidence to show that our Milky Way Galaxy is not the only galaxy in the universe. Before that, no one was sure whether the spiral nebulae, as ***galaxies*** were then called, were just at the rim of our galaxy or were great stellar islands in their own right. The key was improving telescopes to the point where they could detect individual stars in other galaxies, which in turn paved the way to the first rough distance determinations.

FURTHER READING

In the listings below, I offer suggestions for follow-up reading for most of the 50 questions in this book. I have not included technical literature; all of these recommendations are easier reading than, say, *Scientific American*. First, some general references.

The most comprehensive source for general readers is the huge 20-volume series "Voyage Through the Universe," produced between 1988 and 1991 by Time-Life Books. Lavishly illustrated and written in clear, accessible language, it is available in libraries or by personal subscription to the series. For information, write: Time-Life Books, Box C-32068, Richmond, VA 23261-2068. Specific volumes from this collection are mentioned below. They are worth tracking down.

Of the many one-volume general astronomy books available, I immodestly suggest my book *The Universe and Beyond* (Camden House; Camden East, Ontario; rev. ed., 1992), which covers subjects introduced by about half the questions in this book.

Journey Through the Universe by Jay M. Pasachoff (Saunders; New York; 1992) is a beautifully illustrated introductory college text that is very up to date. Another of my favourite references is *Astronomy: The Cosmic Journey* by William K. Hartmann (Wadsworth; Belmont, California; many editions), an outstanding college text that makes a fine reference book.

Q1 THROUGH Q10

All these questions deal with cosmology, the study of the origin, structure and destiny of the universe. A solid introduction to this subject is given by Paul Davies in "Everyone's Guide to Cosmology," *Sky & Telescope*, March 1991. Another useful overview is "The Age Paradox," *Astronomy*, June 1993. Volumes on cosmology in the Time-Life series include *The Cosmos, Workings of the Universe* and *Cosmic Mysteries*. Also recommended: *Atoms of Silence* by Hubert Reeves (MIT Press; Cambridge, Massachusetts; 1984), *The Omega Point* by John Gribbin (Bantam; New York; 1988), *Gravity's Lens* by Nathan Cohen (Wiley; New York; 1988) and *Thursday's Universe* by Marcia Bartusiak (Times Books; New York; 1986). I particularly enjoyed *Lonely Hearts of the Cosmos* by Dennis Overbye (HarperCollins; New York; 1991), a superbly written book about cosmologists and their discoveries. For a dissenter's view, see *The Big Bang Never Happened* by Eric J. Lerner (Random House; New York; 1991).

Q11

Recommended books on the dark-matter enigma include *The Dark Matter* by Wallace and Karen Tucker (Morrow; New York; 1988) and *The Fifth Essence* by Lawrence Krauss (Basic; New York; 1989).

Q13 THROUGH Q15

For more on stars, their nature, sizes, distances, luminosities, and so on, see *Burnham's Celestial Handbook* by Robert Burnham, Jr. (Dover; New York; 1978), a massive but highly readable compilation of cosmic information.

Q16 AND Q17

Distances in space are thoroughly discussed in the

introductory college texts mentioned in the third paragraph of this section.

Q19

The evolution of the sun and other stars is presented in the lavishly illustrated *Cycles of Fire* by William K. Hartmann and Ron Miller (Workman; New York; 1987).

Q20 AND Q21

The background to the discovery of black holes can be found in *Black Holes* by Walter Sullivan (Doubleday; New York; 1979). Contemporary black-hole understanding is presented in *Unveiling the Edge of Time* by John Gribbin (Harmony Books; New York; 1992).

Q22

The concept of wormholes is discussed at length and in entertaining fashion by Paul Halpern in *Cosmic Wormholes* (Dutton; New York; 1992). A more abbreviated account is given by Paul Davies in "Wormholes and Time Machines," *Sky & Telescope*, January 1992.

Q23

For the background to Planet X, the postulated tenth planet, see Chapter 13 of *Planets Beyond* by Mark Littmann (Wiley; New York; 1988).

Q24

A nice update on the search for planets around other stars is "Desperately Seeking Jupiters" by David Bruning, in the July 1992 issue of *Astronomy*. Also try "The Planet Hunters" by Billy Goodman, in *Air & Space Smithsonian* magazine, October/November 1992.

Q25

The case for the Mars "face" is given in excruciating detail in *The Monuments of Mars* by Richard C. Hoagland (North Atlantic Books; Berkeley, California; 1987).

Q26

The New Solar System (Cambridge; New York;

3rd ed., 1990) has extensive material on Jupiter supported by fine illustrations.

Q27

The origin of asteroids is an active area of research, so an up-to-date source is essential. A good start is the asteroids chapter in *The New Solar System* (Cambridge; New York; 3rd ed., 1990) or the asteroids chapter in *Comets, Asteroids, & Meteorites* from the Time-Life "Voyage Through the Universe" series mentioned above.

Q28

Both *The New Solar System* (see Q27 above) and *Moons and Rings* from the Time-Life "Voyage Through the Universe" series provide detailed, well-illustrated information about Saturn's rings and about planetary rings in general.

Q29

One of the astronomical concepts that seem to be consistently underexplained in astronomy books is the orientation of the solar system to the stars and the galaxy. *The Astronomical Companion* by Guy Ottewell has superb illustrations that sort this out. I highly recommend this unique book that is available only directly from the publisher: Astronomical Workshop, Furman University, Greenville, SC 29613. Phone: 803-294-2208.

Q30 AND Q31

A very readable survey of our current knowledge about the moon appears in *Outpost on Apollo's Moon* by Eric Burgess (Columbia; New York; 1993). Also try *The Planetary System* by David Morrison and Tobias Owen (Addison-Wesley; Reading, Massachusetts; 1987).

Q32

The moon illusion is often mentioned in astronomy books but almost always superficially. A clear discussion is in Jay Ingram's *The Science of Everyday Life* (Viking/Penguin; Markham, Ontario; 1989).

Q33

The Helen Sawyer Hogg quotes are from her book

The Stars Belong to Everyone (Doubleday; New York; 1976), unfortunately now out of print.

Q34 AND Q35

Earthshine, haloes and many other atmospheric and astronomical phenomena are described in *Wonders of the Sky* by Fred Schaaf (Dover; New York; 1983). Alan Dyer and I cover similar material in an abbreviated way in Chapter 7 of *The Backyard Astronomer's Guide* (Camden House; Camden East, Ontario; 1991).

Q36

Everything you ever wanted to know about eclipses (and more) is contained in the wonderful book *The Under-Standing of Eclipses* by Guy Ottewell (Astronomical Workshop; Greenville, South Carolina; 1991). See address in Q29 above.

Q37 THROUGH Q44

Practical stargazing issues raised in Chapter 5 are covered in my book *NightWatch* (Camden House; Camden East, Ontario; rev. ed., 1989), which includes colour sky charts, observing tips, an equipment guide and an introduction to astrophotography.

Q45

For the complete story of the Keck telescope, see "The Ultimate Time Machine" by Andrew Chaikin, *Popular Science*, March 1992. Progress on giant telescopes in general is covered in the monthly magazine *Sky & Telescope*. Almost every issue has news about one of the major observatories currently being designed or under construction.

Q46

Details of meteorite strikes in modern times are given in "Against All Odds" by Christopher Spratt, in the March/April 1992 issue of *Mercury*, the journal of the Astronomical Society of the Pacific. Also see "Stop to consider the stones that fall from the sky" by James Trefil, in *Smithsonian* magazine, September 1989.

Q47

The most detailed source of information is *The Star*

of Bethlehem by David Hughes (Walker; New York; 1979). The 2 B.C. Venus-Jupiter conjunction interpretation appeared in *Sky & Telescope*, December 1986.

Q48

UFO Encounters by Jerome Clark (Publications International; Lincolnwood, Illinois; 1992) is a well-illustrated overview of the subject. *UFOs & Outer Space Mysteries* by James E. Oberg (Donning; Norfolk, Virginia; 1982) debunks the "UFOs" supposedly shown in NASA photographs of the moon or seen by astronauts. *Crash at Corona* by Stanton T. Friedman and Don Berliner (Paragon; New York; 1992) documents the persistent allegations that the U.S. military covered up the crash of an alien device in New Mexico in 1947.

Q49

Current NASA plans for SETI are described in *Astronomy*, October 1992, and in *Sky & Telescope*, November 1992. For a readable overview of this subject, see *The Search for Extraterrestrial Intelligence* by Thomas R. McDonough (Wiley; New York; 1987). For an insightful analysis of the SETI question, try the 74-page chapter "Where Are They?" in John L. Casti's superb *Paradigms Lost* (Morrow; New York; 1989). My thoughts on this subject are offered in the final chapter of *The Universe and Beyond*.

Q50

The *Journal of UFO Studies* contains articles from all viewpoints on the UFO question. A special issue on abductions was published in 1989. For information and prices, contact: The J. Allen Hynek Center for UFO Studies, 2457 W. Peterson Avenue, Chicago, IL 60659. For opposite sides of the abduction issue, I suggest *UFO Abductions: A Dangerous Game* by Philip J. Klass (Prometheus; Buffalo, New York; 1989) and *Intruders* by Budd Hopkins (Random House; New York; 1987).

INDEX

THE AUTHOR

Terence Dickinson is Canada's leading astronomy writer. He is the author of 10 books, including the international best-seller *NightWatch*, and writes a weekly astronomy column for *The Toronto Star*. His articles have appeared in more than 50 magazines, ranging from *Reader's Digest* to *Sky & Telescope*. He teaches astronomy part-time at St. Lawrence College, Kingston, Ontario, and is astronomy commentator for *Quirks and Quarks*, CBC-Radio's weekly science programme. Before turning to science writing full-time in 1976, he was editor of *Astronomy* magazine and was an instructor at several science museums and planetariums in Canada and the United States. He has received numerous awards for his work, including the Book of the Year citation from the New York Academy of Sciences and the Royal Canadian Institute's Sandford Fleming Medal for achievements in advancing public understanding of science. He and his wife Susan live in rural eastern Ontario.